Phased-Array Radar Design

Phased-Array Radar Design: Application of Radar Fundamentals

Tom Jeffrey

SciTech
PUBLISHING, INC.

SciTech Publishing, Inc.,
911 Paverstone Drive, Suite B
Raleigh, NC 27615
919) 847-2434, fax (919) 847-2568
scitechpublishing.com.

Editor: Dudley R. Kay

Production Director: Susan Manning

Production Coordinator: Robert Lawless

Cover Design: Kathy Palmisano

Typesetting: J. K. Eckert & Co., Inc.

The publisher and author would like to thank Raytheon for their generous cooperation.

This book is available at special quantity discounts to use as premiums and sales promotions,
or for use in corporate training programs. For more information and quotes, please contact the
publisher.

ISBN: 9781891121692

Library of Congress Cataloging-in-Publication Data

Jeffrey, Tom, 1954–

Phased-array radar design : application of radar fundamentals / Tom Jeffrey.

p. cm.

ISBN 978-1-891121-69-2 (hbk. : alk. paper) 1. Phased array antennas--Design and construction.
2. Radar. I. Title.

TK6590.A6J44 2009

621.3848'3--dc22

2008042350

This book is dedicated to my wife, Marie,
for her continued patience and support

Contents

Preface

This book is intended for systems, hardware, and software engineers who work in the radar field and desire a better understanding of the basics of modern radar design. The book's major objective is to review the fundamental theory and provide examples of its application to the design and analysis of phased-array radars. In general, key theoretical results are presented without derivation or proof except where necessary or useful to understand their application to specific design problems. Ample references are provided throughout the book as sources of both radar theory and other useful information.

INTENDED USES

The book is designed to be used as a self-contained and stand-alone textbook and reference. However, it is assumed that the reader has a basic understanding and familiarity with most undergraduate engineering topics, including calculus, physics, linear systems theory, and probability, along with selected graduate-level subjects such as random processes, digital signal processing, statistical communications, and detection and estimation theory. Ideally, the reader would have the equivalent of a master's degree in electrical engineering, which would include many of these topics.

There are two basic ways to use the book. The first approach is to work sequentially through the chapters. This is appropriate for those who are new to radar or less familiar with applying the theory and desire a systematic treatment, or are using the text as a learner's guide for Raytheon's Advanced Topics in Radar course and follow-on courses. The second approach is to selectively read only those chapters or sections needed to provide specific information. This would be the case for more experienced radar engineers that are interested in refreshing or filling gaps in their existing knowledge.

A central approach used in the book is learning the material via application of the theory to real-world cases studies and example problems. There are several examples interspersed among the chapters for this purpose. These include several aspects of the design of phased-array radars as well as touching on the design of tracking filters and target classification algorithms for air and missile defense missions. In addition, some key mission-related factors associated with the design of air and missile defense radars, as well as missile early warning and surveillance radars, are presented to provide the reader with top-level design guidance for these applications.

This book, when used in conjunction with relevant design courses, study and homework, on-the-job training, diverse radar-related work assignments, and mentoring will enable the reader to become proficient in radar design and analysis. It is also hoped that the importance of careful problem formulation and the trading off of candidate solutions is conveyed. It is expected that these skills and the associated processes will be immediately applicable to design and analysis problems encountered both in academic settings and on the job.

BOOK ORGANIZATION

The book is organized as follows. The first six chapters cover the fundamental radar functions, including detection, waveforms and signal processing, search and acquisition, and target tracking and classification. These topics form the core radar systems theoretic treatment, beginning with a derivation of the general form of the radar range equation and tailored forms for the various radar functions, including volume and horizon search, tracking, as well as for operating in clutter, and jamming environments.

Chapter 7 introduces several data processing functions that are common to most phased-array radars. The key aspects of resource management and radar scheduling are introduced. Other functions treated are radar hardware control, radar echo processing, surveillance, tracking, and target classification. As these are algorithmic in nature and predominantly implemented in real-time software, they are often ignored or only discussed briefly in many radar design textbooks.

Chapter 8 introduces the important capability of interference suppression from sources such as unintentional and intentional radio frequency (RF) emissions. Sidelobe blankers are described, followed by discussions of single and multiple-sidelobe cancellers, open loop nulling, and adaptive processing. The chapter concludes with a description of algorithms such as frequency agility and hopping, and transmit and receive sector blanking, typically software-controlled, which are commonly used interference mitigation techniques.

Chapter 9 introduces phased-array radar architectures. The material is drawn from practical experience rather than theory, as in many of the previous chapters. Five major array-based radar architectures are discussed. These cover full field-of-view (FFOV) and limited field-of-view (LFOV) radars, fixed beam former and digital beam former (DBF) radars, and mechanically steered phased-array radars. Next, the requirements and architectures of narrow and wideband radar architectures are contrasted with one another. Finally, architectures tailored for search, track, fire control, and illumination functions are described.

Chapter 10 addresses a key engineering design tool: the trade study and its application to the design of phased-array radars. After an overview, radar operating frequency selection is examined as a function of mission and desired radar capabilities. Next, the equally important subject of waveform selection is covered for common radar functions (i.e., search, track, and target classification and identification [ID]) in the clear, clutter, jamming, and chaff environments. Next, trades-offs addressing radar coverage and the receiver operating characteristics (ROCs) associated with certain target detection rules are discussed, followed by trades for search, track, and target classification design.

Performance-driven hardware and software requirements are addressed in Chapter 11. The specification of hardware requirements, such as noise figure, phase noise, instantaneous dynamic range, and channel-to-channel alignment are related to specific system-level requirements. Similarly, the data processing requirements associated with tracking, target classification, and signal processing are related to their driving system-level requirements.

The next three chapters address some top-level radar design requirements for the air and ballistic missile defense missions, as well as those for missile early warning systems. The key design drivers for each application are related to the top-level capabilities necessary to perform the particular mission. Like the prior two chapters, this material is also based on my experience in radar system design for these specific types of missions. Chapter 15 examines the task of predicting performance for phased-array radars that execute search, track, target classification, interference suppression, and clutter cancellation functions.

The last section of each chapter provides a list of relevant radar references. These include radar texts, handbooks, and pertinent papers from technical journals. There are also practice problems at the end of some of the early chapters, in addition to the worked examples interspersed within many of the more analytical chapters that serve to illustrate some key concepts.

ACKNOWLEDGEMENTS

This book is based on a number of in-house radar courses I taught at Raytheon over the last twenty-five years. Most the these courses were developed to fill a perceived need by working systems, hardware, and software engineers for basic and more advanced understanding of radar fundamentals necessary to perform their design tasks more effectively.

The book is intended as an application-oriented guide to the design of modern radar systems, primarily ground and ship-based phased-array radars, for which Raytheon is widely acknowledged as a world-class leader. The material is based on both my work and teaching experiences at Raytheon and the General Electric (GE) Company. Over the last many years, I have had the good fortune to work as a radar system engineer on a number of phased-array radars developments, including the BMEWS and PAVE PAWS UHF early warning radars, the family of X-Band ballistic missile defense (BMD) radars that include the THAAD, SBX and High Power Discrimination radars, and the dual S- and X-band radars for the Cobra Judy Replacement shipboard data collection system. Much of the practical experience gained, and the radar design and analysis processes developed, during these assignments are reflected in the book's content.

A number of the radar reference texts cited as references have been used as source material, as well as unpublished notes and teaching materials developed while working at GE and Raytheon. I would like to thank the many talented radar system engineers at both companies who have helped me over the years and whose material in many cases I have reinterpreted and presented here. First, I would like to thank my instructors from the GE Advanced Course in Engineering, who stressed problem-solving skills and the application of basic principles to developing engineering solutions. In addition, I would like to thank Eli Brookner, Fred Daum, Dan Harty, John Krasnakevich, Harry Mieras, Dan Rypysc, and John Toomey of the Raytheon Company for their help and many instructive conversations on radar-related topics and applications during my years with the company. I am also grateful to Dr. Pramod Varshney of Syracuse University for teaching me detection and estimation theory nearly thirty years ago, and to Dr. Yaakov Bar-Shalom of the University of Connecticut, my alma mater, for providing me with a very useful theoretical foundation to develop and apply tracking and data association algorithms over the last twenty-plus years. Finally, I would also like to thank Dan Dechant, who has been a strong supporter of mine as well as a long-term advocate and sponsor, along with Charlene Corey, of systems engineering training at Raytheon.

I would also like to give special thanks to Joe Yu, Dan Bleck, Mike Hart, Tom McDonagh, and Bob Millett, also of the Raytheon Company, without whose thorough review, helpful recommendations, and suggestions this book would not have been published. Last, I want to thank Mel Belcher, of Northrop Grumman Corporation and formerly of Georgia Tech Research Institute, who made many useful suggestions during the early writing process that made this a better book.

Finally, I gratefully acknowledge the good people at SciTech Publishing: Dudley Kay, SciTech's president and my sponsoring editor; Susan Manning, who oversaw all production aspects; Robert Lawless, Susan's production assistant; and cover artist Kathy Palmisano for their encouragement, support, suggestions, and patience as the book went through the important phases of review, layout, proofing, and tying up of the many loose ends. Some unforeseen circumstances slowed completion of the final version of the book, but through it all we arrived at what I believe is the full realization of my goal, and that is the ultimate tribute to a close working relationship between an author and his publisher.

Any errors and omissions are mine, however, and I would be grateful to readers for corrections and suggestions to improve future printings and editions.

Tom Jeffrey
Sudbury, MA
thomas_w_jeffrey@raytheon.com
October 2008

Abbreviations

AAW	Anti-Air Warfare		INS	Inertial Navigation System
ABT	Air-Breathing Target		JPDA	Joint Probability Data
AESA	Active Electronically-Steered Array			Association
			SLB	Sidelobe Blanker
AMTI	Adaptive Moving Target Indicator		SLC	Sidelobe Canceller
			SM-2	Standard Missile-2
ANASIM	Analog Simulation		SMI	Sample Matrix Inverse
AR	Auto-Regressive		SNR	Signal-to-Noise Ratio
ASCS	Antenna Servo Control System		SP	Signal Processor
ATC	Air Traffic Control		STAP	Space-Time Adaptive Processing
A/D	Analog-to-Digital Converter			
BMD	Ballistic Missile Defense		STP	Short-Term Planner
BSC	Beam Steering Controller		STS	Short-Term Scheduler
BSG	Beam Steering Generator		2-D	Two-Dimensional
CDI	Classification, Discrimination, Identification		3-D	Three-Dimensional
			KF	Kalman Filter
CFAR	Constant False Alarm Rate		LFM	Linear Frequency Modulation
CI	Coherent Integration		LFOV	Limited Field-of-View
DA	Data Association		LR	Likelihood Ratio
DBF	Digital Beam Forming		LRT	Likelihood Ratio Test
D-S	Dempster-Shafer		LTP	Long-Term Planner
DIGSIM	Digital Simulation		LTS	Long-Term Scheduler
EKF	Extended Kalman Filter		MFR	Multifunction Radar
EM	Electromagnetic		MHT	Multiple Hypothesis Tracker
EO	Electro-Optics		MHz	Megahertz
ESA	Electronically-Steered Array		MMSE	Minimum Mean Square Error
FFOV	Full Field-of-View		MSLC	Multiple Sidelobe Canceller
FFT	Fast Fourier Transform		MSPAR	Mechanically Steered Phased Array Radar
FOR	Field-of-Regard			
FOV	Field-of-View		MTD	Moving Target Detector
GHz	Gigahertz		MTI	Moving Target Indicator
GLR	Generalized Likelihood Ratio		NB	Narrowband
GLRT	Generalized Likelihood Ratio Test		NCI	Non-Coherent Integration
			NCTR	Non-Cooperative Target Recognition
GPS	Global Positioning System			
IBDA	Innovations-Based Detection Algorithm		NN	Nearest Neighbor
			PAR	Phased Array Radar
ID	Identification		PD	Probability of Detection
IF	Intermediate Frequency		PDA	Probabilistic Data Association
IMM	Interacting Multiple-Model		PFA	Probability of False Alarm

PRF	Pulse Repetition Frequency	SIR	Signal-to-Interference Ratio
PRI	Pulse Repetition Rate	TBM	Tactical Ballistic Missile
RAP	Radar Activity Priority	THAAD	Theater High-Altitude Area
RBF	Receive Beam Former		Defense
RCS	Radar Cross Section	TI	Track Initiation
RF	Radio Frequency	TM	Track Maintenance
RLS	Recursive Least Squares	T/R	Transmit/Receive
RM	Resource Manager	TWS	Track-While-Scan
RMS	Root Mean Square	TWT	Traveling Wave Tube
RRE	Radar Range Equation	UAV	Unmanned Aerial Vehicle
RS	Radar Scheduler	U-D	Upper-Diagonal
SCR	Signal-to-Clutter Ratio	UHF	Ultra-High Frequency
SDP	Signal Data Processor	WB	Wideband

1

Radar Fundamentals

1.1 INTRODUCTION

This chapter introduces some of the major radar concepts that will be described in greater detail in subsequent chapters. One key concept, however, is covered primarily in this chapter: the radar range equation (RRE). Although specific forms of the RRE are discussed in other chapters, its derivation and definitions of all major terms are covered here. The remainder of the chapter provides an overview of other significant radar concepts.

1.2 SEARCH AND TRACK FUNCTIONS

Typically, the functions performed by radars enable the sensing of a target's presence and its physical location and the ability to predict a target's future position. These basic capabilities are fundamental to military radars, such as for fire-control applications, and commercial radars, such as those for air-traffic control (ATC).

The major functions that radars can execute include:

• Search:
  - To survey a volume of space and report locations of targets
  - To measure target position in two or three dimensions (i.e., 2-D or 3-D)

- Tracking:
  - Obtain more accurate target location by "smoothing" measurements
  - Estimate the "state vector" of the target (i.e., positions, rates, and possibly accelerations) to predict the target state vector for a future time

- Track-While-Scan:
  - Combine the search and tracking functions into one radar mode
  - Use data processing to initiate and maintain tracks, while simultaneously searching for new targets
  - Add tracking capability without using additional radar resources.

In order to provide these high-level capabilities, many additional lower-level functions must be executed by radars. Some of these functions are described in the following sections.

1.3 TARGET DETECTION, RESOLUTION, AND CLUTTER CONCEPTS

Three key concepts affecting radar operation are target detection, radar resolution, and clutter backscatter (commonly referred to as simply "clutter"). The following subsections provide descriptions of each and their importance to the operation, capabilities, and performance of different radars.

1.3.1 Target Detection

Radar energy is typically transmitted as a pulse or group of pulses at a carrier or operating frequency ranging between tens and thousands of Megahertz (MHz). The instantaneous bandwidth of each pulse can be on the order of a Megahertz or less (i.e., narrowband operation) or a Gigahertz (GHz) or more (generally considered wideband operation). The reflected energy received by the radar is used to decide whether a target is present or not. Usually this function is performed automatically for targets at all ranges (from the radar) of interest. This is defined as target detection. Target detection is the crucial first function performed by radars. In other words, target detection is a prerequisite for all subsequent radar functions, such as tracking and target classification.

The reflected radar energy (or waveform) is subject to additive thermal noise, primarily due to active electronics in the receive chain. It is in this "noisy" environment within which targets must be sensed. This radar return or "echo" is depicted in Figure 1.1.

Two key target detection attributes are the "probability of detection" (P_d) and "probability of false alarm" (P_{fa}). The radar's design objective is to maximize P_d while maintaining a small (and possibly constant) P_{fa}.

Detectability (or P_d) depends on the ratio of reflected target energy to average thermal noise power, which is defined as the signal-to-noise ratio (SNR). SNR is an important radar performance metric. The performance of nearly all radar functions depends on the SNR. For example, P_d increases monotonically with increasing SNR.

Two approaches to increasing SNR are to use higher-energy (i.e., larger amplitude or longer duration) waveforms or to add (or "integrate") multiple return pulses, either coherently (in-phase) or non-coherently (summing magnitudes or without phase coherence).

A target-present decision when no target (only noise) is present is called a "false alarm." The converse to this error is declaring "no target present" when a target is actually present. This latter error is referred to as a "missed detection."

Multiple pulses can be transmitted as a pulse "train" or "burst." This type of waveform is depicted in Figure 1.2. If each pulse in a burst has energy E, then the waveform illustrated in the figure has total energy NE.

The minimum "non-self eclipsing" range is based on the pulse duration τ and is given by:

$$R_{min} = \frac{c\tau}{2},$$

(1.1)

where c is the speed of light. This phenomenon arises from the fact that reception cannot begin until transmission is complete. The maximum "un-eclipsed" range is therefore defined as:

$$R_{\substack{max \\ uneclipsed}} = c\left(\frac{T-\tau}{2}\right),$$

(1.2)

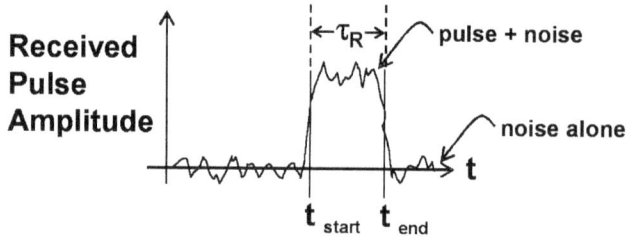

Figure 1.1   Illustration of Received Target Signal or Echo

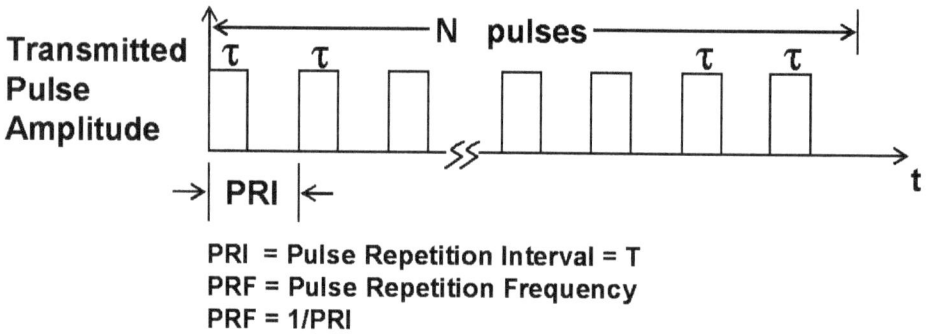

Figure 1.2   A Radar Pulse Train or Burst Waveform

where T is defined as the pulse repetition interval (PRI). Alternatively, the pulse repetition frequency (PRF) is defined as:

$$PRF = \frac{1}{T} = \frac{1}{PRI} \cdot \qquad (1.3)$$

1.3.2   Radar Resolution

When targets are closely spaced in range, angle, or Doppler (proportional to target range-rate) they may not be "resolvable" by the radar. Resolution capability is defined as the minimum separation necessary to declare two targets when there are two targets present. Resolution is a function of waveform duration or bandwidth (range resolution), antenna beamwidth (angular resolution), and coherent integration time (range-rate or Doppler resolution).

A single pulse is a commonly used radar waveform. A simple pulse (envelope) of duration τ in time has a bandwidth, B, approximately given by:

$$B \ = \ \frac{1}{\tau} \cdot \qquad (1.4)$$

Figure 1.3 shows an ideal transmitted rectangular radar pulse. Two targets (of equal amplitude) separated by much greater than τ in time are easily "resolved." The inherent range resolution capability of the waveform is defined as:

$$\delta_R \ = \ \frac{c}{2B} \cdot \qquad (1.5)$$

In real-world applications, to ensure target resolution with a high probability, targets must be separated by 2 to 3 δ_R depending on the targets' relative sizes and the time-sidelobes of the matched filter output, as described in a subsequent section of this chapter.

1.3.3 Clutter Backscatter

The reflected energy received by the radar may originate from unintended objects such as land or sea backscatter, weather (e.g., rain), or man-made objects such as buildings or other structures. This type of radar return is referred to as "clutter." When the radar is used for weather detection, clutter is the desired target.

A key discriminating feature used to differentiate targets from clutter is the observed or measured speed (or range-rate). This method is based on exploiting the Doppler-shifting effect, that is, the phenomenon that returns from higher

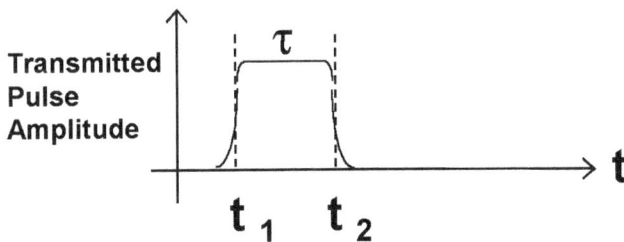

Figure 1.3 Simple Rectangular-Pulse Waveform

range-rate targets of interest are "shifted" in frequency more than lower-speed targets, which in this case are clutter returns. The Doppler frequency-shift based on the target speed is given by:

$$f_D = \frac{2\dot{R}}{\lambda}, \tag{1.6}$$

where $\dot{R}$ is the target or clutter range-rate and λ is the radar operating wavelength.

A second key discriminating feature used to separate targets from clutter is the "polarization" of the reflected electromagnetic (EM) signal. This is especially true for clutter arising from rain. The return from spherical rain drops when using a circularly polarized waveform is reduced compared to the target response when a linearly polarized waveform is employed. Different polarizations will reflect (or refract) distinctly from different shaped targets; for example, sharp edges tend to spread energy into the primary and orthogonal polarizations, whereas smooth reflectors result in a single reflected polarization. Polarization is an important feature for separating certain classes of targets.

1.4 SURVEILLANCE RADARS

Surveillance radars perform an important role for many missions. A surveillance radar's objective is to detect objects in a volume of space and to "acquire" them (i.e., initiate a target track). Usually, these target tracks are handed off to tracking radars or, for multifunction radars, to other radar functions such as tracking or target classification.

A search radar illuminates the volume (i.e., radiates energy into it) with one or more antenna beam positions on transmit. The antenna may be mechanically or electronically scanned across the coverage volume. The "scan" or "frame" time is the time necessary to illuminate the entire volume once.

Most radars use the same antenna to transmit and receive energy. A switch or radio frequency (RF) duplexer is used to direct energy from the transmitter to the antenna or from the antenna to the receiver. The received waveform is very weak relative to the transmitted waveform and must be amplified in the receiver prior to making detection decisions. The receiver also translates the RF energy from its operating frequency to a lower intermediate frequency (IF) or baseband for further processing.

There are a number of different types of searches that these radars can execute. The following subsection describes one of the major search types: volume search. Chapter 4 covers a number of other commonly used searches.

1.4.1 Volume Search

Most surveillance radars that operate in general air search modes, such as in air defense applications (e.g., search and tracking of air targets: aircraft, cruise missiles, unmanned air vehicles), perform volumetric-type searches. As indicated by its name, volume searches survey a volume defined in space, specified by a range extent, and azimuth and elevation extents. A typical volume search beam-pattern (or "raster") employed by a phased-array radar is depicted in Figure 1.4.

A volume search is specified by the physical extents defined in Figure 1.4, the time required to execute the search (i.e., the frame time), the allowed false alarm rate, and the cumulative probability of detection, $P_{d\,cum}$.

1.5 RADAR BLOCK DIAGRAM

Figure 1.5 depicts the block diagram of a basic phased-array radar. As can be seen, the major functions are the phased-array antenna, beam steering generator,

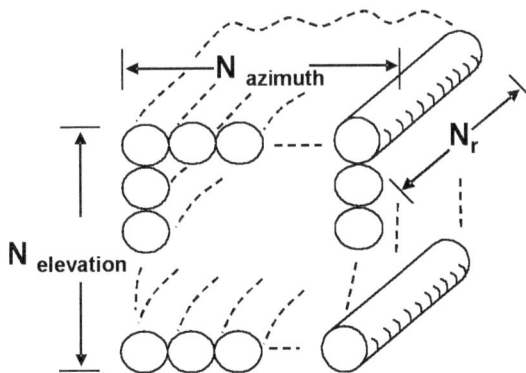

Figure 1.4   Volume Search Beam Raster for Phased-Array Radars

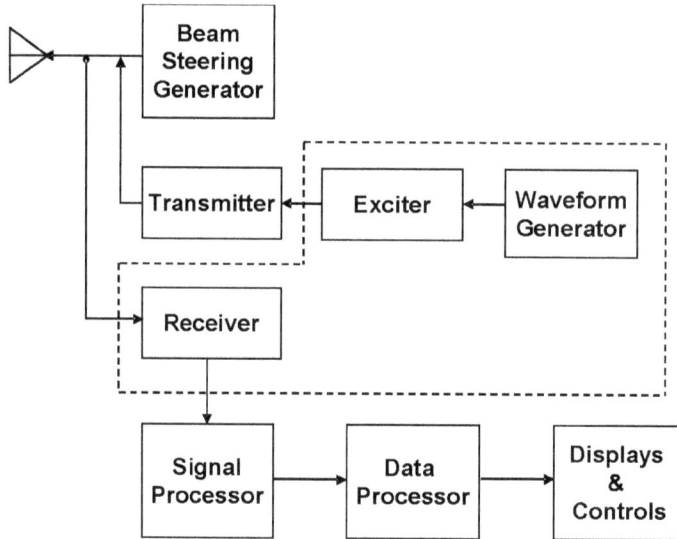

Figure 1.5   Basic Phased-Array Radar Block Diagram

waveform generator, exciter and transmitter, receiver, signal processor, data processor, and operator displays and controls.

Note that for reflector-antenna radars, the phased-array is replaced with a reflector antenna. Also for mechanically scanned radars, the beam steering generator is replaced with antenna pedestal control. In the case of solid-state transmit-receive configuration, the transmitter is replaced with T/R modules (this type of radar is referred to as active-aperture radar). Most often in modern radars, the signal processing is performed in software, and the signal processor and data processor are combined into a single signal/data processor (i.e., computer or multiprocessor).

Figure 1.6 is a block diagram of a phased-array antenna (also known more generally as an electronically-steered array).

A phased-array antenna is a "sampled aperture," where the antenna consists of N small antennas or antenna elements. Each is "phased-steered" to receive or transmit from an angle θ steer using phase shifters. The phase shifters are usually digitally selected to obtain quantized phase (e.g., 0, 22.5°, 45°, 67.5°, 90°). This type of antenna is used for narrowband operation. When wideband operation is required, some combination of time-delay and phase-shift steering is needed.

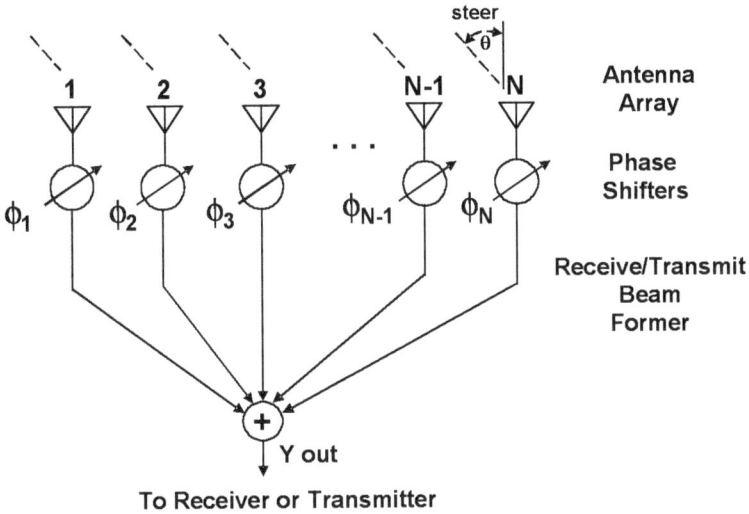

Figure 1.6 Basic Phased-Array Antenna Block Diagram

1.6 RADAR RANGE EQUATION

The radar range equation is the fundamental relationship that defines radar performance for a given set of radar parameters (e.g., peak transmit power, transmit antenna gain, receive antenna gain, wavelength), or the tool used to design a radar to satisfy certain performance requirements (e.g., single-to-noise ratio on a target with a specified radar cross section [RCS] at a specified range from the radar). Figure 1.7 illustrates the basic derivation of the RRE based on the laws of physics such as electromagnetic scattering, EM wave propagation in a vacuum, and so on. As can be seen in the figure, the transmitted power density is defined as:

$$Transmit\ Power\ Density\ =\ \frac{P_t\,G_t}{4\pi\,R^2\,L_t}, \tag{1.7}$$

where $P_t\,G_t$ is the product of the peak transmitter power and the transmit antenna gain, and $R^2\,L_t$ is the product of range of the target from the radar squared and the total transmit losses. For a target with an RCS of σ, the echo or reflected power density is given by:

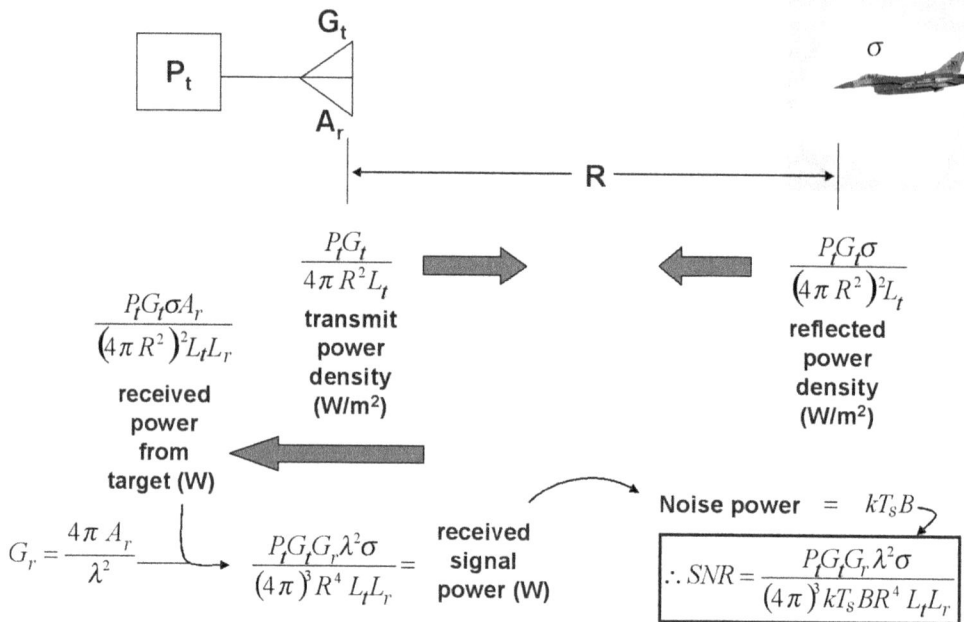

Figure 1.7   Derivation of the Radar Range Equation

$$\text{Echo Power Density} \;=\; \frac{P_t\, G_t\, \sigma}{\left(4\pi\, R^2\right)^2 L_t}\, . \tag{1.8}$$

The received power at the radar antenna aperture of area A_r and receive loss of L_r is then:

$$\text{Power at Radar Aperture} \;=\; \frac{P_t\, G_t\, \sigma\, A_r}{\left(4\pi\, R^2\right)^2 L_t\, L_r}\, . \tag{1.9}$$

When the thermal front-end noise can be modeled as having a "white" spectrum with power spectral density of kT_s, that is, the product of Boltzmann's constant and the system noise temperature, and the radar has a processing (or noise) bandwidth of B, the noise at the output of the matched filter is given by:

$$\text{Noise Power} \;=\; kT_s B\, . \tag{1.10}$$

Therefore, the average signal power-to-RMS noise power, commonly referred to as the SNR, is defined as the ratio of equation (1.9) to equation (1.10), or:

$$SNR = \frac{P_t G_t A_r \sigma}{\left(4\pi R^2\right)^2 kT_s BL_t L_r}. \tag{1.11}$$

If the antenna gain is defined as:

$$G_r = \frac{4\pi A_r}{\lambda^2}, \tag{1.12}$$

then solving for the receive aperture and substituting into equation (1.11) yields the commonly used "sensitivity" form of the RRE:

$$SNR = \frac{P_t G_t G_r \lambda^2 \sigma}{\left(4\pi\right)^3 kT_s B R^4 L_t L_r}. \tag{1.13}$$

The RRE is often calculated in decibel units via:

$$SNR_{dB} = 10 \log_{10} SNR. \tag{1.14}$$

The evaluation of equation (1.14) in tabular form is referred to as a "Blake chart" (named after L. V. Blake). An example of a completed Blake chart for an X-band radar appears in Figure 1.8.

1.6.1   Jamming Effects on Signal-to-Noise Ratio

Equations (1.11) and (1.13) are referred to as the RRE for radars in the "clear" (i.e., operating in a thermal noise environment only). When severe intentional or unintentional interference, commonly referred to as jamming, are present *and* the interference is much larger than the thermal noise, the importance of the SNR is supplanted by the signal-to-interference ratio (SIR) given by:

$$SIR = \frac{S}{I} \approx \frac{S}{N+I} = \frac{1}{\left(SNR\right)^{-1} + \left(SIR\right)^{-1}}. \tag{1.15}$$

PULSE-RADAR-RANGE-CALCULATION WORKSHEET
Based on Skolnick's Radar Handbook, Second Edition, p-2.63

1. Compute System input noise temprature, Ts, following outline in Section A below and Radar Handbook Section 2.5

2. Enter range factors known in other than decibel form in Section B below for reference

3. Enter logarithmic and decibels values in Section C. below, positive values in the plus column, negative values in the minus column. Example: If Do(dB) as given by Figures 2.3 to 2.7 is negative, then -Do(dB) is postive and goes in plus column. For Cb, see Figure 2.1. For definitions of range factors, see Eq. 2.1 and 2.11

Radar antenna height: expressed in feet				Target elevation angle: Θ = 3.00 °		
		h= 50.00 ft				

A. Computation of T_s: $T_s = T_a + T_r + L_r T_e$		B. Range Factors		C. Decibel values	Plus(+)	Minus(-)
		$P_t(kW)$	184.01	10 log10Pt(kW)	22.65	
		$\tau_{\mu s}$	5,000.00	10 log10tms	36.99	
(a.) Compute T_a.		$G_{t(dB)}$	48.90	Gt	48.90	
For $T_{tg} = T_{ta}=290$ and		$G_{r(dB)}$	48.34	Gr	48.34	
$T_g = 36$, use Eq. (2.35b)		$\sigma(m^2)$	0.10	10 log 10 σ		10.00
Read T_a' from Fig. 30.00 deg K		f_{MHZ}	9,500.00	-20 log$_{10}$ f$_{MHz}$		79.55
$L_{a(dB)}$ _____		T_s, °K	484.15	-10 log10 T_s °K		26.85
L_a: 1.00		D_0	15.00	-$D_{0\,(dB)}$		15.00
$T_a = (0.876T_a' -254)/L_a + 290$		C_b		-$C_{b\,(dB)}$		
$T_a =$ 62.28° K		$L_t L_r$	-2.42	-$L_{t\,(dB)}$ -$L_{t\,(dB)}$		2.42
		L_p		-$L_{p\,(dB)}$		
(b) Compute Tr, Equ. 2.37		L_x		-$L_{x(dB)}$		
		Range equation constant 40 log$_{10}$ 1.292			4.45	
Lr(dB): 1.40 Tr= 110.31° K		4. Obtain column totals			161.33	133.82
(c) Compute Te, Eq. (2.38)		5. Enter smaller total below larger			133.82	
		6. Subtract to obtain net decibels			27.51	
Fn(dB) 2.50 Te: 225.70 K		7. Calculate free-space range, Ro, from Ro = 100 x antilog$_{10}$(dB/40) **(NM)**				487.24
Lr: 1.40 LrTe= 311.55 °K		8. Multiply Ro by pattern factor F to obtain				
Add Ts = 484.15° K		non-free-space range R' = RoF. Use Equations (2.41) through(2.56) as appropriate. **F =** 1.00 **RoF=** 487.24				

9. Correct R' for atmospheric attenuation, using Figs. 2.19 through 2.26 as appropriate. Iteration or graphical method may be used. Range Correction Factor for x-dB attenuation is antilog10(x/40)

Two Way Attenuation
L_{atm}(dB) = 1.66
$10^{Latm/40}$ = 0.91   442.94

Bring down results of step 9. This is the   **RADAR DETECTION RANGE (NM)** 442.94
Expressed in kilometers   820.33
RADAR DETECTION RANGE (NM)
Expressed in kilometers

Figure 1.8 Example of a Blake Chart for an X-Band Radar

Therefore, when the interference level is much lower than thermal noise level, the equations (1.11) and (1.13) are a useful performance figure-of-merit (FoM) for radars. However, when the interference level approaches the noise level and exceeds it, the SIR defined in equation (1.15) becomes the performance FoM.

As can be seen from equation (1.15), when the thermal noise is exceeded by the thermal noise-plus-interference, target detection sensitivity is reduced and is now limited by the SIR. Mathematically, this can be expressed as:

$$kT_sB \rightarrow kT_sB + \frac{P_JG_JG_{rj}\lambda^2}{(4\pi R_J)^2} \approx \frac{P_JG_JG_{rj}\lambda^2}{(4\pi R_J)^2}, \tag{1.16}$$

where $P_J G_J$ is the product of the jammer power and antenna gain (referred to as the effective radiated power of the jamming source), R_J is the range of the jammer to the radar, G_{rj} is the radar receive antenna gain in the jammer direction, and jammer bandwidth $\geq B$. Note that the jamming energy impinging on the radar antenna aperture decays as an inverse-range-squared law, compared with the target return, which goes as inverse-range to the fourth power. This enables relatively low-powered jamming sources to degrade radar performance, which is the inherent advantage of the jammer over radars.

Substituting equation (1.16) into (1.13) yields the RRE for the case where jamming levels far exceed thermal noise:

$$SIR = \frac{P_t G_t G_r \lambda^2 \sigma R_J^2}{\left(4\pi\right)^2 P_J G_J R^4 L_t L_r}. \tag{1.17}$$

1.6.2   Other Forms of the Radar Range Equation

There are many forms of the RRE that can be derived from equation (1.13) for specific applications. The RREs used for search and tracking are fundamental variants. These are addressed in the subsequent sections.

1.6.2.1 Radar Range Equation for Volume Search. For volumetric searches as depicted in Figure 1.4, the RRE is adapted in the following manner. First, note that the antenna beamwidth is related to the antenna gain via the relationship:

$$\theta_3 \approx \frac{\lambda}{\sqrt{A}} = \frac{\lambda}{\sqrt{\dfrac{\lambda^2 G}{4\pi}}} = 2\sqrt{\frac{\pi}{G}} . \tag{1.18}$$

The area of a search beam is approximately given by:

$$A_b \approx \theta_{3\,AZ}\, \theta_{3\,EL} = 4\frac{\pi}{G} . \tag{1.19}$$

Also, the average transmitted power can be defined as:

$$P_{AVE} = P_t\, DF = P_t\, \tau\, PRF = \frac{P_t\, PRF}{B} . \tag{1.20}$$

where DF is the radar duty factor (percentage of time allowed to transmit) and PRF is the pulse repetition frequency. Noting that:

$$\frac{\Psi}{\theta_{3\,AZ}\, \theta_{3\,EL}\, T_{SC}} = \frac{\Psi G}{4\pi T_{SC}} = PRF \; (or \; beams \, / \, s) , \tag{1.21}$$

where Ψ is the area to be searched in radians^2 and T_{SC} is the scan or "frame" time for the search, and substituting equation (1.20) into (1.21) yields:

$$\frac{\Psi G}{4\pi T_{SC}} = PRF = \frac{B\, P_{AVE}}{P_t} , \tag{1.22}$$

or:

$$P_t = \frac{P_{AVE}\, 4\pi\, T_{SC}\, B}{\Psi G} . \tag{1.23}$$

Now substituting equation (1.23) into the RRE of equation (1.13) results in:

$$SNR = \frac{P_{AVE}\, G_r\, \lambda^2 \sigma\, T_{SC}}{\left(4\pi\right)^2 k T_s\, R^4\, \Psi L_t\, L_r} = \frac{\sigma\, T_{SC}}{\left(4\pi\right) k T_s\, R^4\, \Psi L_t\, L_r}\, P_{AVE}\, A_r . \tag{1.24}$$

Equation (1.24) is proportional to the $P_{AVE} A_r$ product and is *not* a function of operating frequency. Also note in contrast that equation (1.11), the sensitivity form of the RRE, is proportional to the product $P_t G_t A_r$. Theoretically then, radars at any operating frequency can search a volume equally well if they possess the same $P_{AVE} A_r$.

In practice, however, lower operating frequencies are usually employed for search radars since the number of beams required to search a given volume, for a fixed antenna aperture size, is many fewer than at higher frequencies. At higher frequencies this can lead to timeline occupancy issues since the larger number of beams required will demand that longer portions of the timeline be scheduled. When these scheduling periods exceed the specified frame time, the radar is said to be "occupancy-limited."

1.6.2.2 Radar Range Equation for Horizon Fence Search. Early warning (EW) missile surveillance radars and ballistic missile defense (BMD) radars often employ horizon fence searches to detect and acquire targets. Figure 1.9 depicts a typical search fence that covers ± 60 degrees of azimuth.

The fence concept is based on the fact that if radars have adequate detection range on the targets of interest, then any ascending ballistic target must fly through a fence and will be detected. Therefore, instead of performing a radar resource-intensive volume search (i.e., due to the much larger number of beams), a single row of beams at or above the horizon is sufficient for missile search and acquisition applications.

Starting with the RRE of equation (1.24) for volume search, modifications can be made to take advantage of the horizon fence characteristics. First note that for missiles flying through the fence:

$$T_{SC} = \frac{\theta_3}{\dot{E}_T} = \frac{\theta_3 R}{N v_T},$$
(1.25)

where $\dot{E}_T$ is the target elevation rate, R is the target range, v_T is the vertical target velocity, and N is the number of looks required for detection. Noting the beamwidth relationship in equation (1.18), (1.25) can be expressed as:

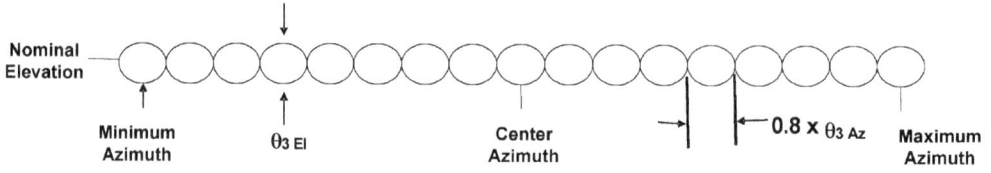

Figure 1.9 Typical Horizon Fence Search Pattern

$$T_{SC} = \frac{2R}{N v_T} \sqrt{\frac{\pi}{G}} \cdot \qquad (1.26)$$

Substituting equation (1.26) into (1.24) yields:

$$SNR = \frac{\sigma}{\left(2\sqrt{\pi}\right) k T_s R^3 \Psi N v_T L_t L_r} \frac{P_{AVE} A_r}{\sqrt{G_r}} \cdot \qquad (1.27)$$

As can be seen from equation (1.27), the RRE for horizon fence search is weakly a function of operating frequency due to the $\sqrt{G}$ term and is also follows an inverse R^3 law rather than R^4 as do the RREs in equations (1.13) and (1.24).

1.6.2.3 Radar Range Equation for Tracking. The primary driving requirement on tracking is angle accuracy given by:

$$\sigma_\theta = \frac{\theta_3}{k_m \sqrt{2\ SNR\ \eta}} , \qquad (1.28)$$

where k_m and η are the monopulse slope and number of independent measurements smoothed by the tracking filter. The SNR is defined by the track sensitivity form of the RRE given by equation (1.11). Substituting average power for peak power as defined in equation (1.20) yields:

$$SNR = \frac{P_{AVE}}{PRF} \frac{G_t A_r \sigma}{\left(4\pi R^2\right)^2 k T_s L_t L_r} , \qquad (1.29)$$

Substituting equation (1.29) into the square of equation (1.28) yields:

$$\sigma_\theta^2 = \frac{\theta_3^2}{2\,k_m^2\,\eta}\,\frac{PRF\left(4\pi\right)^2 R^4\,k\,T_s\,L_t\,L_r}{P_{AVE}\,A_r\,G_t\,\sigma}.$$

(1.30)

Now substituting for antenna beamwidth from equation (1.18) and noting that η = $PRF\,T_t$ where T_t is the time in track (and PRF refers to the track update rate) yields:

$$\sigma_\theta^2 = \frac{\left(4\pi\right)^3}{2\,k_m^2\,T_t}\,\frac{k\,T_s\,R^4\,L_t\,L_r}{P_{AVE}\,A_r\,G_t\,G_r\,\sigma}.$$

(1.31)

As can be seen, the tracking accuracy is inversely proportional to $P_{AVE}\,AG^2$, or equivalently $P_{AVE}\,A^3/\lambda^4$, and is therefore highly dependent on operating frequency. For a given size antenna aperture, superior tracking accuracy is achieved by higher-frequency radars.

1.6.2.4 Radar Range Equation Summary. Tables 1.1 and 1.2 provide a summary of the RREs for the radar applications discussed in this section.

1.7 DETECTION IN NOISE

This topic is covered in detail in Chapter 2 for detection in noise, clutter, and jamming environments, respectively. However, some of the basic concepts are introduced in this chapter.

In general, detection performance is a function of SNR (or in cases of clutter or jamming, signal-to-clutter ratio (SCR) or SIR, respectively) and typically employs threshold tests to declare target detections. References [6] and [7] address the theory for this critical radar function. Most modern radars use matched-filter receivers to maximize the SNR at the processed output prior to detection decisions.

Table 1.1 Forms of the Radar Range Equation (Square Antenna)

Radar Application	Appropriate Form of the Radar Range Equation	Radar Parametric Factor
Volume search	$SNR = \dfrac{\sigma\, T_{SC}}{\left(4\pi\right) k\, T_s\, R^4\, \Psi\, L_t\, L_r}\, P_{AVE}\, A_r$	$P_{AVE}\, A$
Horizon fence search	$SNR = \dfrac{\sigma}{\left(2\sqrt{\pi}\right) k\, T_s\, R^3\, \Psi\, N\, v_T\, L_t\, L_r}\, \dfrac{P_{AVE}\, A_r}{\sqrt{G_r}}$	$\dfrac{P_{AVE}\, A}{\sqrt{G}}$
Track sensitivity	$SNR = \dfrac{P_t\, G_t\, A_r\, \sigma}{\left(4\pi\, R^2\right)^2 k\, T_s\, B\, L_t\, L_r}$	$P_t\, A\, G$
Track accuracy	$\sigma_\theta^2 = \dfrac{\left(4\pi\right)^3}{2 k_m^2\, T_t}\, \dfrac{k\, T_s\, R^4\, L_t\, L_r}{P_{AVE}\, A_r\, G_t\, G_r\, \sigma}$	$P_{AVE}\, A\, G^2$

Table 1.2 Forms of the Radar Range Equation (Circular Antenna)

Radar Application	Appropriate Form of the Radar Range Equation	Radar Parametric Factor
Volume search	$SNR = \dfrac{\sigma\, T_{SC}}{16\, k\, T_s\, R^4\, \Psi\, L_t\, L_r}\, P_{AVE}\, A_r$	$P_{AVE}\, A$
Horizon fence search	$SNR = \dfrac{\pi\, \sigma}{16\, k\, T_s\, R^3\, \Psi\, N\, v_T\, L_t\, L_r}\, \dfrac{P_{AVE}\, A_r}{\sqrt{G_r}}$	$\dfrac{P_{AVE}\, A}{\sqrt{G}}$
Track sensitivity	$SNR = \dfrac{P_t\, G_t\, A_r\, \sigma}{\left(4\pi\, R^2\right)^2 k\, T_s\, B\, L_t\, L_r}$	$P_t\, A\, G$
Track accuracy	$\sigma_\theta^2 = \dfrac{\left(2\pi\right)^4}{2 k_m^2\, T_t}\, \dfrac{k\, T_s\, R^4\, L_t\, L_r}{P_{AVE}\, A_r\, G_t\, G_r\, \sigma}$	$P_{AVE}\, A\, G^2$

The basic form of the detectors is:

$$s(t) \overset{H_1}{\underset{H_0}{\begin{array}{c} > \\ < \end{array}}} \begin{array}{c} V_T \\ V_T \end{array} , \tag{1.32}$$

where $s(t)$ is the output of the optimal matched filter and V_T is the detection threshold. The detection threshold is usually based upon an assumption of Rayleigh-distributed random noise magnitudes.

1.7.1 Target Models

Chapter 2 describes the performance of equation (1.32) for several analytical target models. For the purposes of this chapter, it is assumed that targets follow an exponentially distributed random variation in power (equivalent to a Rayleigh-distributed random fluctuation voltage model). This is referred to as a Swerling I target model when returns are independent look-to-look (i.e., scan-to-scan rather than pulse-to-pulse).

Detection is modeled as a statistical phenomenon with two sources of statistical or non-deterministic variability:

- Additive interference (e.g., noise, interference)
- Fluctuating target radar cross section.

The target's RCS is an approximation devised to account for the portion of scattered energy reflected back toward the radar from the target. The target RCS may vary in amplitude (or fluctuate) versus time. The RCS depends on the RF operating frequency of the radar (e.g., ultra-high frequency [UHF], L-band, X-band) and on the reflecting structure of the particular target. Complex scattering surfaces such as an aircraft are composed of many individual RF scatterers. Simpler target shapes, like a conical missile warhead, may present only one or two scattering sources.

The interpretation of the reflected energy scattered by the target then becomes driven by the waveform range resolution:

$$\delta_R = \frac{c}{2B} = \frac{c\tau_{eff}}{2} . \tag{1.33}$$

When δ_R is $\geq$ target "length," a single return occurs at the radar that is a sum of the complex-valued individual scattering components. RCS fluctuations arise due to constructive and destructive combination of individual scattering center components within the resolution-limited waveform bandwidth.

However, when δ_R << target "length," multiple scattering returns are resolved in range (resulting in little or no RCS fluctuation). Models of target RCS fluctuation (e.g., Swerling Models I-IV, log-normal) were developed to allow detection analysis to be performed without having actual RCS data versus viewing angle.

1.7.2   Detection and False Alarm Probabilities

The probability of false alarm affects the number of "false target responses" that must be evaluated by the radar processor. If a search volume such as that in Figure 1.10 consists of N_b antenna beams, where $N_b = N_{azimuth} \times N_{elevation}$, and N_r range gates, then the average number of false alarms per frame or search volume is:

$$N_{FA} = P_{FA} \cdot N_b \cdot N_r \cdot \tag{1.34}$$

Typically, the search P_{FA} is selected to result in $\leq N_{FA}$/second. Therefore, for example, if N_{FA}/second = 5 false alarms/second, and $N_b = 100$ and $N_r = 1000$, then for a scan or frame time (T_{SC}) of 2 seconds, the P_{FA} allowed is:

$$P_{FA} = \frac{N_{FA} \cdot T_{SC}}{N_b \cdot N_r} = \frac{(5)(2)}{10^5} = 10^{-4} \cdot \tag{1.35}$$

Estimation of radar detection performance can be obtained analytically using RCS models such as the Swerling models. Each model type has an associated probability density (or family of probability densities) than can be integrated to calculate P_D:

$$P_D = \int_{V_T}^{\infty} f_X(x)\, dx , \tag{1.36}$$

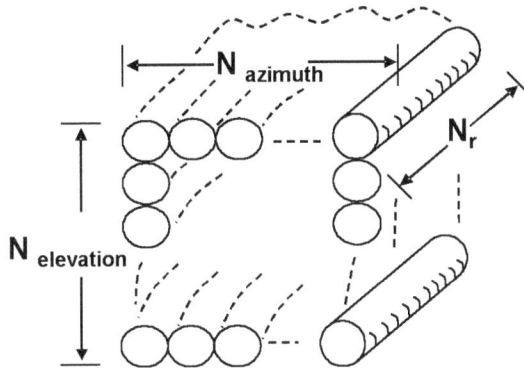

Figure 1.10   Example of Search Volume

where $f_x(x)$ is the RCS fluctuation density, and V_T is the selected detection threshold, and:

$$P_{FA} \quad = \quad \int_{V_T}^{\infty} f_n(n)\, dn \quad = \quad e^{-V_T^2 / 2\sigma^2} \, , \tag{1.37}$$

where the threshold can be calculated to be:

$$V_T^2 = \left(-\ln P_{fa} \right) 2\sigma^2 \, . \tag{1.38}$$

Substituting (1.38) into (1.36) allows the calculation of P_D for any analytical RCS model.

1.7.3   Detection in Thermal Noise

Calculating P_D for a specified P_{FA} and target RCS model requires performing the integration defined in (1.36). For exponentially distributed RCS (e.g., Swerling I), the integral has a closed-form solution:

$$P_D \quad = \quad \left(P_{FA} \right)^{\frac{1}{1 + SNR}} \, . \tag{1.39}$$

Equivalently, for a Swerling I target fluctuation (slow fluctuation—scan-to-scan), the required SNR is given by:

$$SNR_{REQ} = \frac{\ell n\, P_{FA}}{\ell n\, P_D} - 1.$$

(1.40)

For probability densities in (1.36) that do not have a closed-form solution, numerical integration can be used to evaluate P_D. P_D has been evaluated for many target models and is documented in texts such as references [6] and [7].

1.7.4 Constant False Alarm Rate Processors

The detection threshold defined in equation (1.38) is function of two parameters, P_{FA} and σ^2. The first is specified as described in Section 7.2. The noise power, σ^2, is established by the receiver noise figure or system noise temperature (T_s) that appears in the denominator of the expression for the RRE. Accurately estimating the noise level, which is itself a random process, can be challenging depending on the radar's operating environment.

A constant false alarm rate (CFAR) processor is a method to estimate σ^2 in a localized manner around the range cell to be tested for detection. A typical CFAR block diagram is illustrated in Figure 1.11. There are many variants of CFAR algorithms. These are described in more detail in Chapter 3. A very common type is the cell-averaging (CA) shown in the figure. The basic concept is to use two relatively short (in terms of the number of range cells) sliding windows before (i.e., lagging) and after (i.e., leading) the cell-under-test (CUT) to be evaluated as a candidate for detection to estimate the noise power.

Each of these sliding windows can provide an independent estimate of σ^2, denoted $\hat{\sigma}^2$, with the estimation error variance being a function of the number of cells employed in the estimate. These estimates can be based on simple arithmetic averages, with and without censoring (i.e., removing large contributors like adjacent targets or range sidelobes from consideration). Independent of the precise formula used, the resulting estimate is used to compute:

$$V_T^2 = -2\,\hat{\sigma}^2 \ln P_{FA}.$$

(1.41)

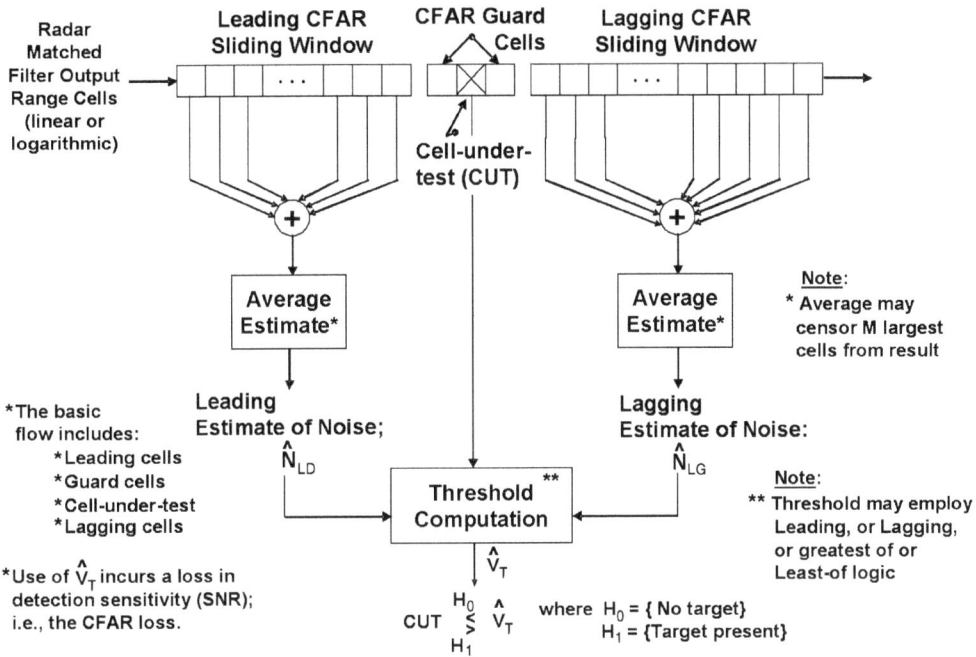

Figure 1.11 Block Diagram of a CFAR Processor

Due to the variance on the total noise estimate, use of equation (1.41) in place of equation (1.38) incurs a loss in detectability referred to as a CFAR loss. For this reason, CFAR tends to be used where the additional signal processing loss (included in the denominator of the RRE as a component of L_r) can be tolerated. Therefore, it is usually used for tracking but not for functions such as long-range search where the additional loss is undesirable.

1.7.5   Detection in Clutter

A target can be detected in the presence of clutter (unwanted RF backscatter) if there is some distinguishing feature (or features) that allows separation of the two "echoes." Examples of features include:

• Doppler shift (range-rate difference)

• Polarization

where:

$$\Delta f_d = \frac{2\,\Delta \dot{R}}{\lambda},\tag{1.42}$$

and $\Delta \dot{R}$ and λ are the difference between the target or clutter range-rates and the radar operating wavelength, respectively. A typical case is illustrated in Figure 1.12.

For example, land clutter is usually distinguishable from moving targets since the mean velocity of the land reflections is zero. However, Doppler shifts for rain in wind conditions or sea clutter at high sea state can overlap with those of slow-moving targets (especially surface targets like ships or tanks). As a benefit to target detection, certain RF waveform polarizations, for example, circular, can result in reduced reflections from rain clutter as compared with those of the target.

There are two basic approaches to mitigating clutter effects to better detect moving targets:

• Moving Target Indicator (MTI) Cancellers

• Pulse-Doppler Processing.

MTI cancellers operate by subtracting sequential radar returns with the objective that near-stationary clutter will be cancelled, but moving targets will not.

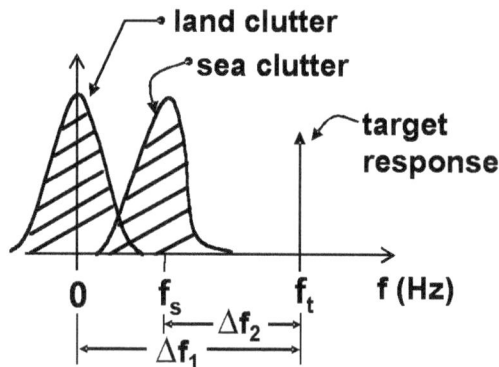

Figure 1.12 Doppler Frequency Separation of Target from Land and Sea Clutter

The simplest MTI canceller is the single-delay (or two-pulse) canceller depicted in Figure 1.13. The clutter appearing at the canceller output is reduced substantially by the attenuation of low-frequency clutter. Optimum performance is obtained when target Doppler shift is near PRF/2 (i.e., least attenuation of target response).

Pulse-Doppler waveforms are coherent bursts that consist of N pulses spaced by a uniform delay of PRI (pulse repetition interval). A pulse-Doppler pulse train waveform is shown in Figure 1.14.

The matched filter to a Pulse-Doppler waveform consists of sub-pulse matched filtering (i.e., range processing), with the results stored for M range cells, followed by N Doppler filtering. Clutter (and its "alias") will appear in lower and higher Doppler filters. This is depicted in Figure 1.15.

1.8 RESOLUTION AND MEASUREMENT ACCURACY

Resolution is defined (for any measurement) as the target separation distance necessary to identify when two targets are present. Waveform resolution is defined as the inherent separation capability of the waveform in range and/or

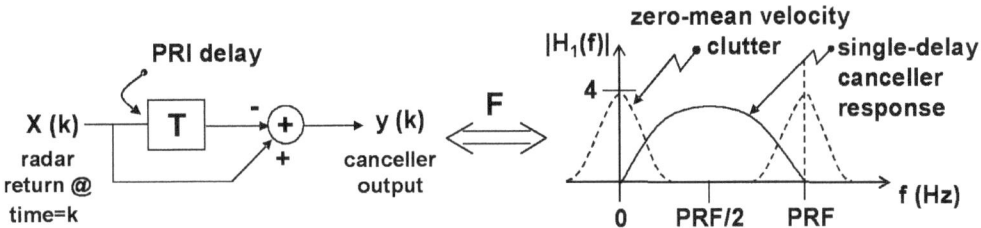

Figure 1.13 Single-Delay MTI Canceller and Associated Frequency Response

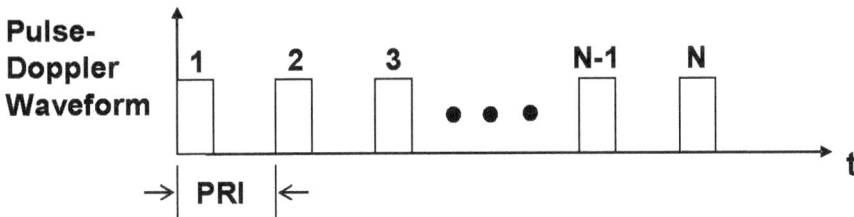

Figure 1.14 Pulse Train Waveform Used for Pulse-Doppler Processing

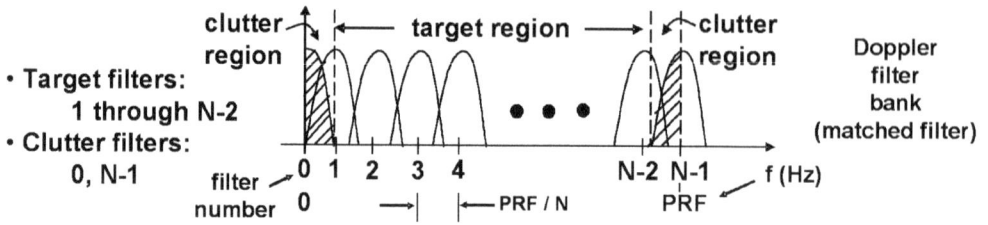

Figure 1.15 Pulse-Doppler Processing and Associated Frequency Response

Doppler. Angular resolution is defined as the inherent angle separation capability of the sum antenna pattern.

Definitions of resolution include:

$$
\begin{aligned}
Range: \quad & \delta_R = \frac{c}{2B} \\
Angle: \quad & \delta_\theta = \theta_3 \\
Doppler: \, & \delta_f = \frac{1}{T}
\end{aligned}
\qquad\qquad (1.43)
$$

where:

   τ = pulse length
   B = pulse bandwidth
   c = speed of light
   θ_3 = 3 dB antenna beamwidth
   T = integration time.

Some convenient rules of thumb include the following:

• Two-target separation capability for equal amplitude targets at "reasonable" signal-to-noise ratio, is approximately 2δ, where δ is the relevant inherent resolution

• For unequal target amplitudes, with "reasonably high" matched-filter sidelobes, the approximate resolution can be in the range: 2.5δ–3δ.

Measurement accuracy can be modeled as a function of the inherent resolutions: δ_R, δ_θ, and δ_f as:

$$Range\ accuracy \quad \cong \quad \frac{\delta_R}{\sqrt{2\ SNR}} = \sigma_R$$

$$Angle\ accuracy \quad \cong \quad \frac{\delta_\theta}{k_m\sqrt{2\ SNR}} = \sigma_\theta \;\; ;\; k_m \approx 1.6\,. \qquad (1.44)$$

$$Doppler\ accuracy \quad \cong \quad \frac{\delta_f}{\sqrt{2\ SNR}} = \sigma_f$$

As can be seen, for this model accuracy improves in an inversely proportional manner to $\sqrt{SNR}$ where SNR is the signal-to-noise ratio at the output of the radar matched filter. This model is based on the approximate statistical "Cramer-Rao bound" to the estimation error standard deviation for each parameter.

The basic range resolution of an "un-coded" pulse of duration τ is given by δ_R = $c\tau/2$, where c = speed of light. In general, since large values of τ correspond to higher SNRs, resolution for this waveform is inversely proportional to detectability.

Pulse compression was developed to counteract this phenomenon. By "coding" the simple pulse (modulating the carrier over the pulse duration), good resolution can be achieved without sacrificing SNR by increasing the effective bandwidth (B). Two popular approaches to achieving pulse compression are:

• Linear frequency modulation waveforms (LFM, also known as "chirp" waveforms)

• Discrete phase-coded waveforms (digital modulation).

These two techniques are illustrated in Figure 1.16.

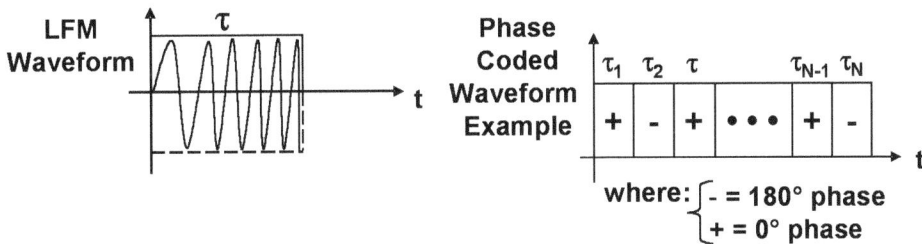

Figure 1.16 Linear Frequency Modulation and Phase-Coded Waveforms

1.9 TRACKING RADARS AND THE MONOPULSE TECHNIQUE

Tracking radars are typically used to obtain and maintain improved accuracy compared to a search or surveillance radar. This capability is usually achieved using a combination of hardware and software techniques, including:

- Monopulse antenna processing for angle measurements
- Wider RF bandwidth to improve range resolution and accuracy
- Pulse-Doppler waveforms to improve range-rate resolution and accuracy (and for clutter mitigation)
- Signal processing techniques, such as range interpolation, to improve range accuracy
- Tracking filters to improve accuracy of position and rate estimates and predictions.

Tracking radars predict the future position of targets by estimating range and angular rates. These predictions are used to position the antenna beam for subsequent radar transmit and receive actions.

Monopulse is a technique to measure target angles by separating two-dimensional receive antennas into azimuth and elevation "quadrants"; and combining them in a specific way to enable the estimation of angular positions. An "error" pattern is formed as $e(\theta) = (\theta)/\Sigma(\theta)$ as shown in Figure 1.17. The amplitude and sign of $e(\theta)$ indicates the distance of the target from the antenna pointing angle.

1.10 TRACK-WHILE-SCAN RADARS

Track-While-Scan (TWS) is an approach that combines the search and track functions:

- While searching a volume, target acquisition (verification and track initiation) and track maintenance are performed using detections
- No dedicated tracking beams are scheduled, but the "normal" search scan is used for both search and track functions.

For mechanically scanned radars, like air-traffic control (ATC) radars:

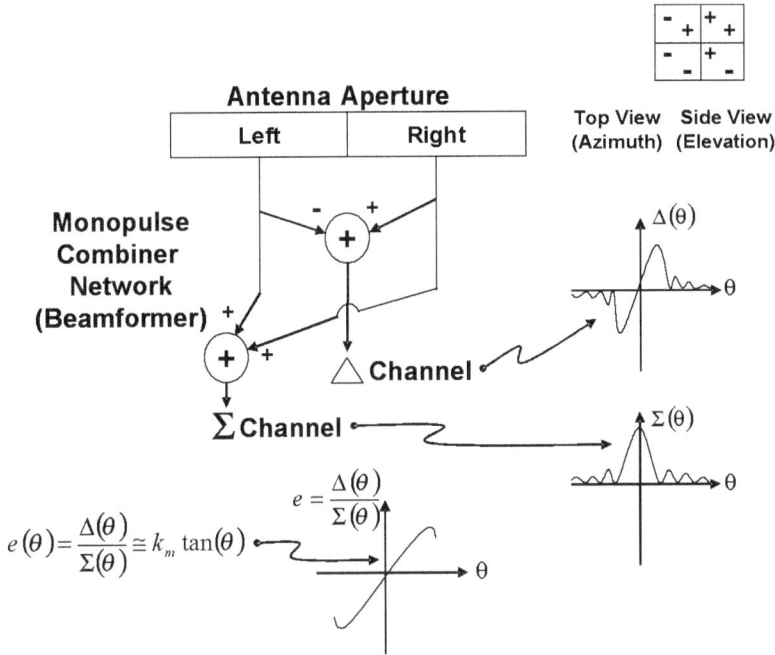

Figure 1.17   Illustration of Monopulse Antenna Processing

- A rotating antenna performs a periodic scan over 360 degrees of azimuth

- A broad elevation beam is typically used to achieve maximum elevation target illumination.

For phased-array radars, similar search volumes (e.g., raster-scanned discrete antenna beams) are serviced, again both for search and track purposes:

- Usually, the previous two or three scans of data are used to perform track initiation (TI)

- Subsequently, tracks are maintained (or updated) on each new scan (re-visit) of a particular angular region.

1.11 REFERENCES

[1] R. Nitzberg, *Radar Signal Processing and Adaptive Systems,* Artech House, 1999
[2] D. K. Barton, *Modern Radar System Analysis,* Artech House, 1988
[3] D. R. Wehner, *High Resolution Radar,* Artech House, 1987

[4] Y. Bar-Shalom & X. Li, *Multitarget-Multisensor Tracking*, YBS, 1995
[5] S. Haykin & A. Steinhardt, *Adaptive Radar Detection and Estimation*, Wiley-Interscience, 1992
[6] H. Van Trees, *Detection, Estimation and Modulation Theory, Part 1*, Wiley-Interscience, 2001
[7] J. DiFranco & W. Rubin, *Radar Detection*, SciTech, 2004

1.12 PROBLEMS

1. Consider a radar with a peak power of 100 kW, 45 dB transmit antenna gain, a 0 dBsm target, a 10 square meters antenna aperture, a target range of 500 km, with transmit and receive losses of 3 dB each. Estimate the received power at the radar aperture using the relationship below.

$$P_{rec} = \frac{P_t G_t \sigma A_r}{\left(4\pi R^2\right)^2 L_t L_r} .$$

2. Consider a radar being degraded by barrage noise jamming. For a radar receiver with system noise of –143 dBm, a jammer effective radiated power ($P_J G_J$) of 10 W, a 0.01 square meter effective radar antenna aperture (equivalent to a –30 dB sidelobe), with the jammer at 1,000 km:

$$N_{effective} = kT_s B_r + \frac{P_J G_J A_r'}{4\pi R_J} .$$

How much degradation in equivalent noise power does the jammer produce? Using the radar from problem 1, estimate the effective signal-to-interference ratio (SIR). Is this good or bad for the radar's performance?

3. Consider a radar with 2 false alarms per second, 1,000 range cells, 16 beams, and a 10-second search frame time. What probability of false alarm does this require? What if the search requires 32 beams and a 5-second frame time?

$$P_{FA} = \frac{N_{FA} \cdot T}{N_b \cdot N_r} .$$

4. Consider the probabilities of false alarm from problem 3. If the SNR achieved by the radar is 15 dB, then calculate the corresponding probabilities of detection assuming a Swerling I RCS fluctuation.

$$P_D = \left(P_{FA}\right)^{\frac{1}{1+SNR}} \qquad SNR_{REQ} = \frac{\ell n\, P_{fa}}{\ell n\, P_d} - 1$$

5. If a probability of detection of 0.95 is required for the system, what SNRs are required to achieve these for the probabilities of false alarm calculated in problem 3? Are these SNRs reasonable for a radar? Why or why not?

6. Consider a radar with range, angle (both azimuth and elevation), and Doppler resolutions of 15 meters, 20 milliradians, and 100 Hz. If an SNR of 10 dB is available, calculate the measurement accuracies that can be expected. If the original target was a fighter aircraft with a 10 dBsm RCS, what accuracies would be achieved for a missile with a 0 dBsm RCS and for a 747 aircraft with a 25 dBsm RCS? Are these reasonable?

$$Range\ accuracy \cong \frac{\delta_R}{\sqrt{2\ SNR}} = \sigma_R$$

$$Angle\ accuracy \cong \frac{\theta_3}{k_m\sqrt{2\ SNR}} = \sigma_\theta$$

$$Doppler\ accuracy \cong \frac{\delta_f}{\sqrt{2\ SNR}} = \sigma_f$$

7. If an LFM or "chirp" waveform with a 10 μsec pulse length and a 1,000 MHz bandwidth is used, what is the effective range resolution? What range accuracies would be achieved for the three target types in problem 5?

2

Target Detection

2.1 INTRODUCTION

This chapter addresses an important aspect of radar design: the detection of targets in real-world environments. The theory of radar detection is covered thoroughly in many text books, including references [2] through [5]. The chapter's main objective is to describe the practical application of target detection concepts to radar design and analysis.

First, target detection is described as the determining the presence of the "desired" object in different environments, including:

- "Undesired" thermal noise
- Reflections from natural and man-made objects (i.e., clutter).

The commonly used Swerling target models are described, and their applicability is discussed. Next, target detection in both surface and volume clutter is described. Last, multiple-pulse detection methods are described, including coherent and non-coherent integration.

2.2 Target Radio Frequency (RF) Scattering Models

Real targets are not usually "point scatterers" of incident radio frequency (RF) energy (i.e., point targets), but are a complex combination of individual scatterers dependent on the radar viewing geometry, RF operating frequency, and bandwidth. For very wide modulations bandwidth radars (e.g., B ≥ 500 MHz), the radar can usually resolve individual RF scattering centers on the target's physical configuration. As long as these resolved scatterers are not obscured from the radar (e.g., via blockage or geometry), they will generally appear to the radar as nonfluctuating echoes. Narrowband radars, however, will *not* resolve individual scatterers. This situation leads to the constructive and destructive combination of scattering sources, and ultimately results in target fluctuations. Figure 2.1 illustrates these concepts pictorially for an aircraft target.

Figure 2.1 addresses the concept of RF scattering from air targets encountered by air defense radars. Scattering centers arise due to specular points and discontinuities on a target's surface when illuminated by a short radar pulse. Returns from an air target may result from reflections of the nose, cockpit, wings, engine cavities, and engine turbine blades of an aircraft as depicted in Figure 2.2. As stated previously, the composite target response will be a function of whether or not these scatterers are resolved by the radar's waveform.

The appropriate target scattering models are a function of physical structure (e.g., size, shape, scattering centers, and reflected polarizations), radar operating frequency, and waveform and processing parameters (e.g., bandwidth, integration time).

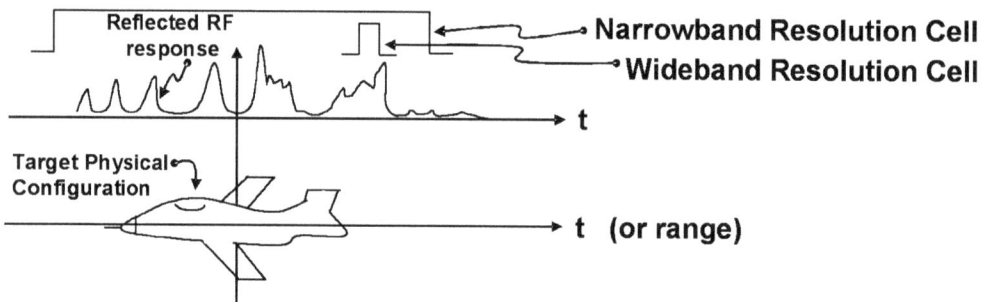

Figure 2.1 Wideband Target Scattering versus Radar Resolution

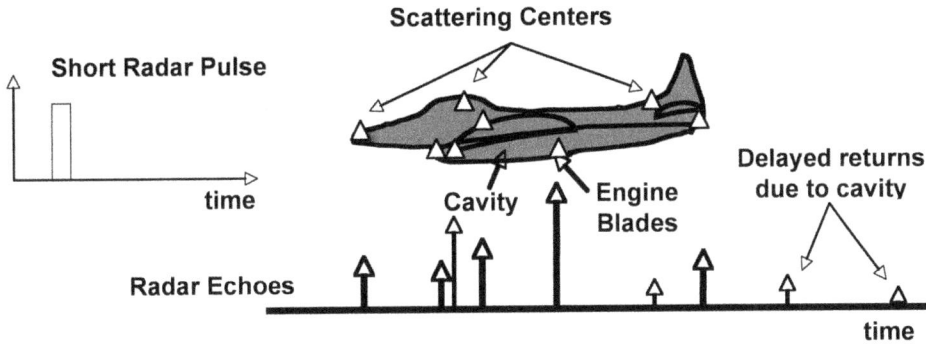

- Scattering centers are specular points and discontinuities on the target surface
- Polarization and amplitude of a peak depend on incident polarization, scattering center shape and orientation - response is a function of target shape
- All peaks vary slowly in amplitude and position as target aspect to radar changes

Figure 2.2   RF Scattering from Air Targets

Air targets are in general complex scattering sources that consist of many individual scattering centers as indicated in Figure 2.2. Alternatively, ballistic missile target objects encountered by missile defense radars are typically simpler targets. Figure 2.3 depicts two such targets, a re-entry vehicle (RV) or warhead and an attitude control module (ACM) or post-boost vehicle (PBV).

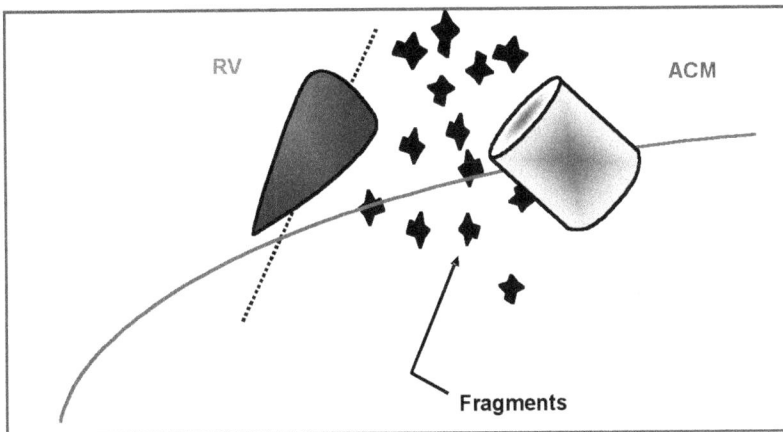

Figure 2.3   Representative Ballistic Missile-Shaped Targets

Simulated radar cross sections (RCSs) for two tumbling ballistic missile-shaped objects are shown in Figure 2.4 as a function of time. As can be seen, there is a periodic structure to the amplitudes of the radar returns due to differences in target scattering as the objects tumble relative to the radar line-of-sight (LOS).

2.3 TARGET DETECTION IN NOISE

This section reviews some fundamental detection theory. For a nonfluctuating target (i.e., constant amplitude), the return radar echo is of amplitude A and duration T. The return in each radar (resolution) cell is compared to a fixed threshold voltage set above background thermal noise. When the return amplitude exceeds the threshold a "target" is declared. Otherwise, "no target" is declared per the test:

$$r(t) \underset{\substack{< \\ H_0}}{\overset{\substack{H_1 \\ >}}{}} V_T \quad , \tag{2.1}$$

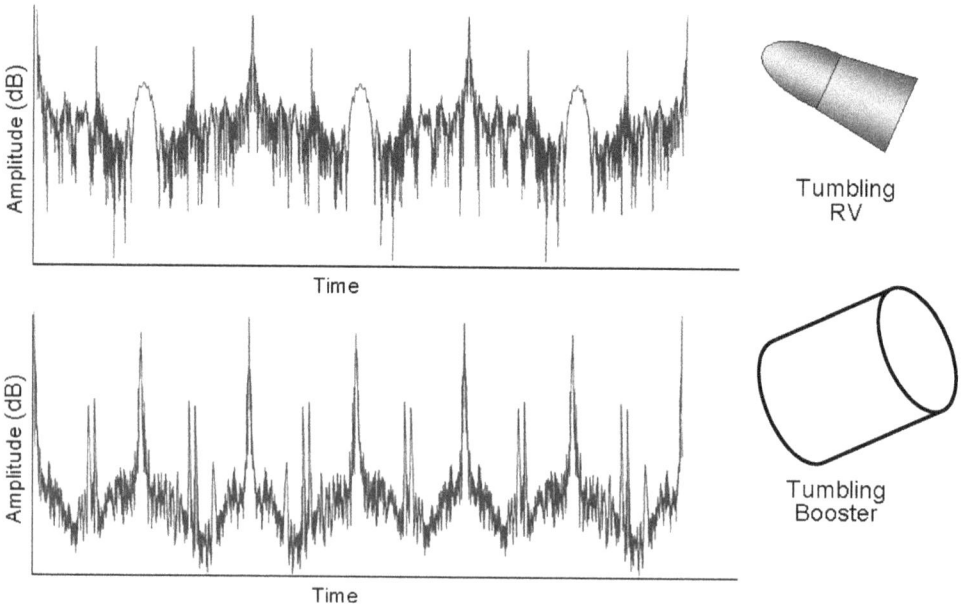

Figure 2.4 Simulated Return Amplitudes for Tumbling RV and Booster-Shaped Objects

where $r(t)$, V_T, H_0, and H_1 are the radar return from the target, the detection threshold, and the hypotheses for no target (i.e., the null hypothesis) and for target a target present, respectively.

The probability of detection (P_D) for the nonfluctuating target is a function of the signal-to-noise ratio (SNR):

$$ SNR \ = \ \left(\frac{A^2}{2} \right) \left(k T_s B \right)^{-1} \ = \ \left(A^2 \big/ 2 \right) N_0 \, B \ = \ \left(A^2 T \big/ 2 \right) \Big/ N_0 \, , \qquad (2.2) $$

where T_s, N_0, and B are the system noise temperature, noise power, and noise bandwidth, respectively. The target detection problem is depicted graphically in Figure 2.5.

Prior to detection, at the quadrature detector, the target voltage is:

$$ V_t(t) \ = \ A_I(t) \cos \left(\omega_c t \right) \ + \ A_Q(t) \sin \left(\omega_c t \right) , \qquad (2.3) $$

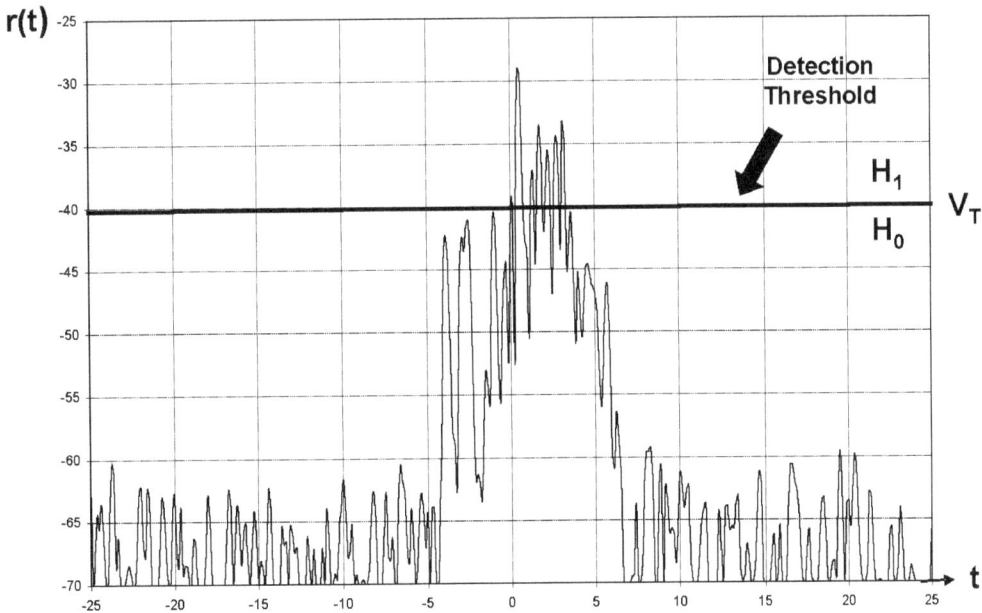

Figure 2.5 Illustration of the Target Detection Problem

where A_I, A_Q, and ω_c are the in-phase and quadrature waveform envelopes and radian carrier frequency, respectively. The noise voltage at the quadrature detector is:

$$V_n(t) \;=\; X_I(t)\cos(\omega_c t) + Y_Q(t)\sin(\omega_c t)\cdot \tag{2.4}$$

The signal-plus-noise complex-envelope is then:

$$\begin{aligned}
r(t) &= \left\{\left[A_I(t)+X_I(t)\right]^2+\left[A_Q(t)+Y_Q(t)\right]^2\right\}^{\frac{1}{2}} \\
&= \left[X^2(t)+Y^2(t)\right]^{\frac{1}{2}}
\end{aligned} \tag{2.5}$$

If the noise statistics are bi-variable Gaussian (with zero-mean), the joint probability density in r and θ (electrical phase angle) is:

$$f(r,\theta) \;=\; r\,\exp\!\left(A_p^2/2\sigma^2\right)\exp\!\left[-\left(r^2-2r\left(A_x\cos\theta+A_y\sin\theta\right)\right)/2\sigma^2\right], \tag{2.6}$$

where:

$$\begin{aligned}
A_p^{\,2} &= A_x^{\,2}+A_y^{\,2} \\
2\sigma^2 &= \textit{Total Noise Power}
\end{aligned} \tag{2.7}$$

By integrating over θ, the density for r becomes:

$$f_r(r) \;=\; r\,\exp\!\left(-A_p^2/2\sigma^2\right)\exp\!\left(-r^2/2\sigma^2\right)I_o\!\left(rA_p/\sigma^2\right)r/\sigma^2, \tag{2.8}$$

where I_0 is the Bessel function of the first kind. The P_D is given by the integral of (1) from V_T (threshold) to ∞:

$$P_D \;=\; \int_{-V_T}^{\infty} f_r(r)\,dr \;=\; Q\!\left(A_p/\sigma, V_T/\sigma\right)\cdot \tag{2.9}$$

where Q is the Marcum Q-function. Equation (2.9) can be evaluated for a specific SNR and detection threshold, V_T. Usually, an acceptable probability of false alarm, P_{FA}, is specified. For the Rayleigh noise distribution, the P_{FA} is given by:

$$P_{FA} \quad = \quad \int_{-V_T}^{\infty} \frac{\alpha}{\sigma^2} e^{-\left(\frac{\alpha}{\sigma}\right)^2 \big/ 2} \, d\alpha \quad = \quad e^{-\frac{V_T^2}{2\sigma^2}} . \tag{2.10}$$

Solving for the detection threshold yields:

$$V_T^{\ 2} \quad = \quad \left(-\ell n \, P_{FA}\right) 2\sigma^2 . \tag{2.11}$$

For a Swerling I radar cross section fluctuation model:

$$f_\gamma\left(\gamma\right) \quad = \quad \frac{1}{\overline{SNR}} \exp\left(-\frac{\gamma}{\overline{SNR}}\right), \tag{2.12}$$

where: $\qquad\qquad \overline{SNR} = Average \ SNR$.

$$P_{D_{SWI}} \quad = \quad \int_{V_T}^{\infty} f_\gamma\left(\gamma\right) d\gamma \quad = \quad \int^{\infty} f_\gamma\left(\gamma\right) d\gamma \quad = \quad \left(P_{FA}\right)^{\frac{1}{1+\overline{SNR}}} . \tag{2.13}$$

The Swerling I fluctuating target model assumes an exponential RCS (or power) probability density; or equivalently, a Raleigh voltage probability density. Another possible target model exhibits RCS fluctuations that follow a chi-square distribution with 4 degrees-of-freedom:

$$f_\gamma\left(\gamma\right) \quad = \quad 4\left(\frac{\gamma}{\overline{SNR}^2}\right) \exp\left[-\left(\frac{2\gamma}{\overline{SNR}}\right)\right]. \tag{2.14}$$

This is the Swerling III model. As to the applicability of these models, note that:

- Swerling I is appropriate for slowly fluctuating (e.g., scan-to-scan for search) RCS dominant point targets
- Swerling III is more appropriate for slowly fluctuating dominant multiple-scattering center targets.

Integrating the Swerling III density from V_T to ∞ yields:

$$P_D \atop SW\,III = \left[\frac{2}{\left(2+\overline{SNR}\right)^2}\right]\left[2+\left(\left(V_T/\sigma\right)^2/2\right)\overline{SNR}+\overline{SNR}^2/2\right]$$
$$\exp\left[-\left(V_T/\sigma\right)^2/2/\left(1+\overline{SNR}/2\right)\right] \qquad (2.15)$$

The benefit of Swerling III (SW III) over Swerling I (SW I) is primarily when non-coherently integrating pulses at higher SNRs. SW III is superior to SW I for detection purposes when $SNR \geq 10$ dB.

Some detailed Swerling model definitions and characteristics are:

- For two N-pulse bursts, 1 per scan:
  - Return amplitudes from N pulses on Scan A are statistically independent of amplitudes from N pulses on Scan B (i.e., {A1, A2, ... , AN}, {B1, B2, ... , BN})

- Swerling I & III assume Set of A amplitudes have identical value, and Set of B amplitudes have identical value: $A1 = A2 = ... = AN$ and $B1 = B2 = ... = BN$

- Swerling I assumes $f\gamma\,(\gamma)$ from equation (2.12) and Swerling III assumes equation (2.14) applies

- Swerling II and IV assume each reflection in Sets A & B are statistically independent, and that equation (2.12) and (2.14) apply, respectively: $A1$ not equal to $A2$, etc., and $B1$ not equal to $B2$, etc.

In summary, Swerling I & III are slowly fluctuating targets (i.e., scan-to-scan, search frame-to-frame). Swerling II & IV assume fast RCS fluctuations (i.e., pulse-to-pulse). Swerling I & III can be converted to Swerling II and IV, respectively, if pulse-to-pulse RF changes are sufficient to de-correlate the target of interest (a function of target shape, etc.).

2.4 TARGET DETECTION IN CLUTTER

The previous section discussed target detection in thermal noise. In this section the effects of backscatter from natural and man-made objects will be modeled as an augmentation to a thermal noise background. Doppler processing is a key technique for mitigating clutter, since in general, the targets of interest are moving faster than the apparent clutter velocity. Two basic classes of waveforms, with different matched filter (MF) processors, are:

• Moving target indicator (MTI) waveforms and processing (cancellers)

• Pulse-Doppler (PD) waveforms and processing (Doppler filter banks).

Clutter discriminants are features that allow separation of targets from clutter returns. Candidate clutter discriminants include:

• Elevation

• Velocity

• Polarization

• Carrier frequency sensitivity

• Azimuth angle sensitivity

• Signal bandwidth

Each of these is discussed next.

Target elevation: Most often, target returns are from above "ground level" and land clutter returns are not; therefore, elevation is useful for separating targets from ground clutter. Target velocity: The most commonly used feature for separating targets from ground, sea, and weather clutter. These clutter types are generally stationary to slowly moving. Since most targets of interest exhibit higher radial speeds, velocity can be a useful clutter discriminating feature. Target polarization sensitivity: Certain RF polarizations reflect from targets differently than from undesired objects (e.g., rain backscatter). An example is the use of circular polarization, which can suppress rain clutter relative to target objects by up to 10 dB. Carrier frequency sensitivity: Certain RF frequencies maximize target-to-clutter responses, and use of frequency diversity can maximize target-to-clutter ratio. RF bandwidth: Higher RF bandwidth reduces range cell dimension (and increases range resolution) and therefore reduces area (sea, land) and volume (rain, chaff) clutter backscatter.

The primary quantity of interest in characterizing clutter magnitude is the equivalent radar cross section of the clutter, since the signal-to-clutter ratio (SCR) (for main beam clutter at the target range) is given by:

$$\left(S/C\right) = \frac{RCS_{t\,arg\,et}}{RCS_{clutter}}.$$

(2.16)

Two basic types of clutter must be addressed: (i) area clutter (land, sea), and (ii) volume clutter (rain, chaff). The defining relationships for these two clutter types are provided by the effective radar cross section of the clutter:

$$\sigma_C = \begin{cases} \sigma^\circ A_C \\ \sigma^\circ V_C \end{cases},$$

(2.17)

where σ°, A_C, and V_C are the normalized clutter coefficient, area of the surface clutter, and volume of the volumetric clutter, respectively. The units of σ° are (m^2/m^2) for surface clutter and (m^2/m^3) for volume clutter to result in the clutter RCS having units of m^2.

Clutter area and volume are defined as:

$$A_C = \left(\frac{Rc\tau}{2}\right)\tan\phi\,\theta_{AZ}$$

(2.18)

and:

$$V_C = R^2\theta_{AZ}\theta_{EL}\left(c\tau/2\right),$$

(2.19)

where R, c, τ, and ϕ are the clutter range, speed of light, pulse length, and depression angle, respectively. θ_{AZ} and θ_{EL} are the antenna azimuth and elevation beamwidths.

In many cases of interest, clutter is neither in the antenna main beam nor at the target range. For this more general case, the SCR is derived from the radar range equation for received power defined as:

$$P_{rec} = \frac{P_t G_t \sigma A_r}{\left(4\pi R^2\right)^2 L_t L_r} ,$$ (2.20)

where σ is the effective RCS of either the target or clutter. For surface clutter, the signal-to-clutter ratio is defined as the ratio of the received target power to the received clutter power:

$$SCR = \frac{P_t G_t \sigma A_r^2}{\left(4\pi R^2\right)^2 L_t L_r} \frac{\left(4\pi R_c^2\right)^2 L_t L_r}{P_t G_t \sigma_{clutter} A_r^{'2}}$$

$$SCR = \left(\frac{A_r}{A_r^{'}}\right)^2 \left(\frac{R_c^3}{R^4}\right) \left(\frac{\sigma}{\sigma^{\circ}\left(\frac{c\,\tau}{2}\right)\tan\phi\ \theta_{AZ}}\right) ,$$ (2.21)

where R_c, R, A_r, and A'_r are the range to the clutter, the range to the target, the full antenna aperture, and the effective antenna aperture of the sidelobe upon which the clutter impinges, respectively.

Since velocity (or more precisely range-rate or radial speed) is a key clutter discriminant, the Doppler frequency shift of the RF carrier frequency is usually exploited to mitigate clutter returns' effect on target detection. Two major classes of Doppler techniques are:

- MTI waveforms and MTI delay line cancellers
- Pulse-Doppler (PD) waveforms and Doppler filter banks.

Both types will be discussed. However, in brief, the difference between the two methods is:

- MTI:
  - Excellent performance against stationary narrowband clutter; simple processing that uses two to three pulses per detection
- Pulse Doppler:
  - Excellent performance against stationary or moving clutter. Uses coherent burst (N-pulse) waveform and a bank of N-Doppler filters for detection.

Moving target indicator waveforms and processing exploit the low-speed nature of land, and sea and weather clutter. A typical MTI application uses three
pulses (two delays), and the processing shown in Figure 2.6.

The response of the three-pulse (two-delay) MTI canceller shown is given by
Figure 2.7.

Clutter attenuation (CA) of an MTI canceller is defined as:

$$CA = \frac{C_{in}}{C_{out}},$$ (2.22)

where C_{in} and C_{out} are the input and output clutter powers, respectively.

The Doppler-frequency dependent improvement factor, $I(f_D)$, is given by:

$$I(f_D) = \frac{(Signal-to-Clutter)_{out}}{(Signal-to-Clutter)_{in}}.$$ (2.23)

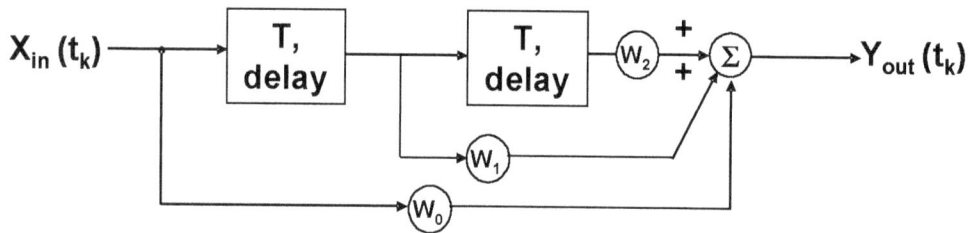

Figure 2.6   Three-Pulse MTI Processor

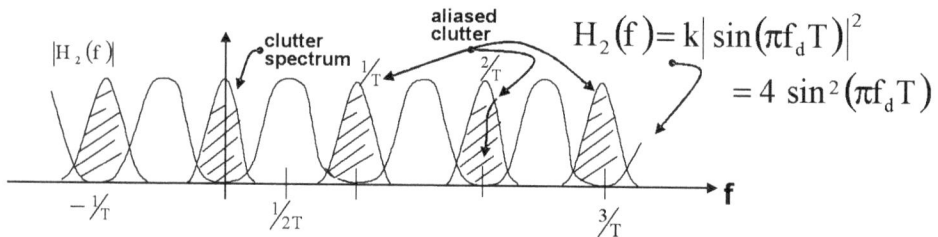

Figure 2.7   MTI Canceller Frequency Response

and:

$$I(f_D) = \frac{(Signal)_{out}}{(Signal)_{in}} CA \quad (for\ a\ linear\ receiver). \tag{2.24}$$

The average improvement factor (averaged over frequency) for a two-pulse canceller is defined as the clutter attenuation and is given by:

$$I_2 = CA \approx \left[2(\pi\sigma_f T)^2\right]^{-1} = \left(PRF/\sigma_f\right)^2 / 19.75, \tag{2.25}$$

where σ_f, T, and PRF are the clutter spectral width, pulse repetition interval (PRI), and pulse repetition frequency, respectively. For the three-pulse canceller, the CA is given by:

$$I_3 \approx \left(\frac{PRF}{\sigma_f}\right)^4 / 780. \tag{2.26}$$

The generalized MTI canceller transfer function is given by:

$$H_M(f) = g_M\left[1 - \exp(-j2\pi f_D T)\right]^M \tag{2.27}$$

or:

$$H_M(f) = g_M \sum_{m=0}^{M} (-1)^m \binom{M}{m} \exp(-jm2\pi f_D T); \quad \binom{M}{m} = binomial\ coefficient \tag{2.28}$$

The weights for $M = 1, 2, 3,$ and 4 are:

$$\begin{aligned}
M &= 1 \Rightarrow (1, -1) \\
M &= 2 \Rightarrow (1, -2, 1) \\
M &= 3 \Rightarrow (1, -3, 3, -1) \\
M &= 4 \Rightarrow (1, -4, 6, -4, 1)
\end{aligned} \tag{2.29}$$

These correspond to $M = 1$: single-delay or two-pulse canceller; $M = 2$: double-delay or three-pulse canceller; $M = 3$: triple-delay or four-pulse canceller; and $M = 4$: quadruple-delay or five-pulse canceller. For all cases, the sum of weights equal to zero ensures a null in the response at zero Doppler. For many applications, a three-pulse canceller with PRF stagger will satisfy clutter cancellation requirements.

When clutter is nonstationary, such as sea clutter at high sea states, or rain clutter in medium to high wind, the mean Doppler shift is well above zero frequency. For these types of clutter, MTI cancellers are not effective since the clutter spectrum is not centered in the canceller notch.

An alternative waveform and processing to MTI cancellers is the pulse-Doppler (PD) waveform (an N-pulse coherent burst) processed using a bank of Doppler filters after the range matched filter. PD waveforms have several advantages over the simpler MTI waveforms, and associated cancellers, including:

• Targets receive close to the theoretical optimum SNR gain relative to noise (N^2 versus N)

• PD waveform processing using a bank of Doppler filters can deal with non-zero or frequency-offset Doppler clutter

• More degrees of freedom are available to tailor the effective transfer function, H(f), when a bank of Doppler filters are used.

The simplest implementation of a Doppler filter bank is the output of FFTs (fast Fourier transforms [FFTs]) in the Doppler domain, computed from the range-cell-sampled, matched-filter output sub-pulses of the pulse train. A narrowband target will "integrate up" in one (or two or three) Doppler filters. Typically, one Doppler filter will contain the majority of the target energy. Figure 2.8 illustrates the PD waveform and its processing.

The advantages of PD waveforms and Doppler processing have been described (i.e., coherent target gain, mitigates nonzero Doppler clutter). However, the specific "post processing" associated with Doppler processing needs to be described.

First, some form of automatic detection processing is desirable. A constant false alarm rate (CFAR) processor is usually employed for this purpose. The con-

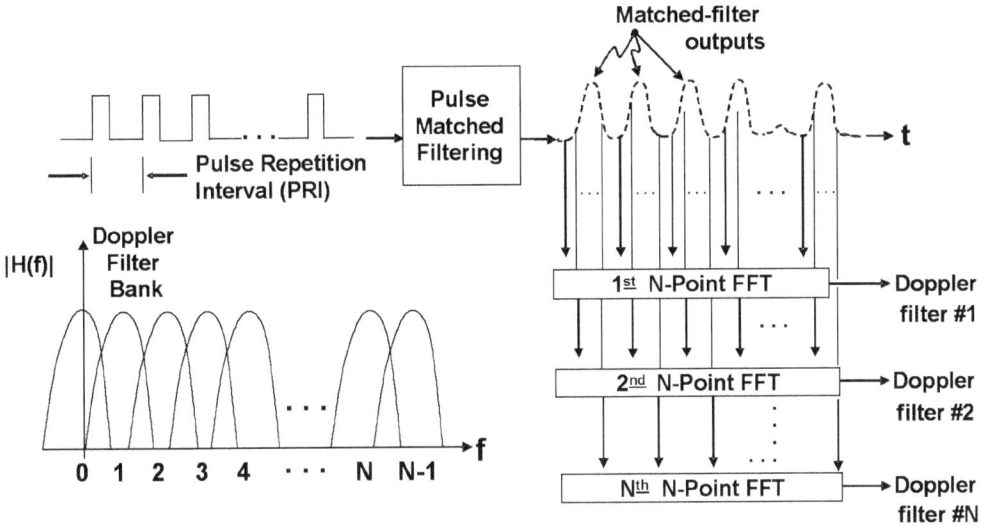

Figure 2.8 Pulse-Doppler Waveforms and Processing

figuration typically used is a CFAR processor subsequent to each Doppler filter. Next, the filters dominated by clutter must be identified and ignored for subsequent targets return candidacy. Automatic methods to detect clutter are based upon CFAR background estimation, as well as a prior knowledge of likely filters (frequency or "speed" likelihood) that might contain clutter (e.g., using a clutter map).

Typically, the zero and $N-1^{th}$ filters are dominated by stationary clutter, especially at low altitudes. Moving clutter (e.g., sea clutter at higher sea states, blowing rain, etc.) will usually occur in the lower and (symmetrically) higher Doppler filters.

2.5 MULTIPLE-PULSE DETECTION

The previous section addresses the detection of targets in clutter using multiple pulses and either the MTI or pulse-Doppler signal processing approach. However, it is often necessary to integrate multiple pulses to enhance the SNR to an adequate level for detecting weak targets or targets at long slant ranges. This section addresses three multiple-pulse detection techniques: (i) binary (or M-out-of-N) integration, (ii) non-coherent integration, and (iii) coherent integration.

2.5.1 Binary Integration

This technique uses multiple pulses to increase detection probabilities by defining a detection criterion of detecting at least M times out of N opportunities. This technique enhances detectability at the slight cost of increased probability of false alarm. When this technique is employed, the effective probability of detection is given by:

$$P_{D\ M-out-of-N} = \sum_{m=M}^{N} \left(\frac{N!}{(N-m)!\,m!} \right) (P_D)^m (1 - P_D)^{N-m}. \tag{2.30}$$

For the special case of $M = 1$, equation (2.30) reduces to:

$$P_{D\ M-out-of-N} = 1 - (1 - P_D)^N. \tag{2.31}$$

Since there are N opportunities for a false alarm when using this technique, the probability of false alarm for M-out-of-N detection is given by:

$$P_{FA\ M-out-of-N} = N\,P_{FA}. \tag{2.32}$$

2.5.2 Non-Coherent Integration

A second type of multiple-pulse detection is non-coherent integration. This technique adds pulses in a root-mean-square (RMS) sense before applying the threshold test in equation (2.1). This approach can be beneficial when adding pulses for fluctuating targets at reasonable SNRs as compared to the coherent integration approach described in the next section. In addition, since no phase information is used in this technique, non-coherent integration can be used in conjunction with frequency diversity which, as discussed in Section 3, can improve detectability for Swerling target models.

Due to the RMS addition of target returns in non-coherent integration, the effective SNR of the integrated return is approximately given by:

$$SNR_{NCI} \approx \sqrt{\sum_{i=1}^{N} SNR_i}, \tag{2.33}$$

where is the SNR of the ith return from the transmitted pulse train. When the SNRs of each return are approximately equal, equation (2.34) becomes:

$$SNR_{NCI} \approx \sqrt{N}\ SNR_1 , \qquad (2.34)$$

where SNR_1 is the SNR of each return. Equation (2.34) is the rule of thumb for the expected benefit of non-coherent integration relative to single pulse detection. Now, since the noise components of each return are added in an RMS sense, the effective P_{FA} is given by:

$$P_{FA\ NCI} = \sqrt{N}\ P_{FA\ 1} , \qquad (2.35)$$

where $P_{FA\,1}$ is the probability of false alarm for each return.

2.5.3   Coherent Integration

A third type of multiple-pulse detection is coherent integration. This technique adds pulses as voltages with knowledge of phase before applying the threshold test in equation (2.1). Equivalently, coherent integration can be considered to be vector addition of the multiple returns. However, since phase information is used in this technique, coherent integration cannot be used in conjunction with frequency diversity, which destroys coherency from return-to-return. In addition, this type of integration is limited by the coherency time which is the shorter of the times that the radar hardware and target remain coherent. In most cases, the target correlation time constant is the limiting component.

Due to the vector addition of target returns in coherent integration, the effective SNR of the integrated return is approximately given by:

$$SNR_{CI} \approx \sum_{i=1}^{N} SNR_i , \qquad (2.36)$$

where SNR_i is the SNR of the ith return from the transmitted pulse train. When the SNRs of each return are approximately equal, equation (2.34) becomes:

$$SNR_{CI} \approx N\ SNR_1 , \qquad (2.37)$$

where SNR_1 is the SNR of each return. Equation (2.37) is the rule of thumb for the expected benefit of coherent integration relative to single pulse detection. Since the noise components of each return are again added in an RMS sense for coherent integration, the effective P_{FA} is given by:

$$P_{FA\ NCI} = \sqrt{N}\ P_{FA\ 1},$$ (2.38)

where P_{FA} is the probability of false alarm for each return. This is the same as for non-coherent integration.

2.6 REFERENCES

[1] A. Papoulis, *Probability, Random Variables, and Stochastic Processes*, McGraw-Hill, 1965
[2] D. K. Barton, *Modern Radar System Analysis*, Artech House, 1988
[3] D. K. Barton, *Radar System Analysis and Modeling*, Artech House, 2004
[4] H. Van Trees, *Detection, Estimation and Modulation Theory, Part 1*, Wiley-Interscience, 2001
[5] J. DiFranco & W. Rubin, *Radar Detection*, SciTech, 2004
[6] E. Brookner, *Aspects of Modern Radars*, Artech House, 1988
[7] N. Levanon, *Radar Principles*, Wiley-Interscience, 1988
[8] R. Nitzberg, *Radar Signal Processing and Adaptive Systems*, 2nd Edition, Artech House, 1999
[9] M. Skolnik, *Introduction to Radar Systems*, 3rd Edition, McGraw-Hill, 2002
[10] M. Skolnik, *Radar Handbook*, 2nd Edition, McGraw-Hill, 1990
[11] S. Haykin & A. Steinhardt, *Adaptive Radar Detection and Estimation*, Wiley, 1992

2.7 PROBLEMS

1. Consider a radar that employs a fixed-noise threshold to achieve a specified probability of false alarm. Calculate the threshold-to-noise ratio to achieve false alarm probabilities of 10^{-M} where $M = 3, 4,$ and 5. Note that:

$$\left(\frac{V_T^2}{2\sigma^2}\right) = -\ln P_{FA},$$ (2.39)

where $2\sigma^2$ is the noise power at the output of the matched filter.

2. Consider a radar being degraded by barrage noise jamming. For a radar receiver with system noise (kTB) of –143 dBm, a jammer effective radiated power ($P_J G_J$) of 10 W, a 0.01 square meter radar antenna aperture (equivalent to a –30 dB sidelobe), with the jammer at 1,000 km, calculate the necessary thresholds for the probabilities of false alarm from problem 1. If the radar has 15 dB SNR in the clear (i.e., no jamming) at a specified target slant range, calculate the detection probabilities with and without jamming for a Swerling I target model for these values of P_{FA}. Remember that:

$$N_{effective} = kT_s B_r + \frac{P_J G_J A_r^{'}}{\left(4\pi R_J\right)^2}.$$

3. Calculate the probability of detection for a Swerling III target with nominal SNRs of 10, 15, and 20 dB, respectively, for $P_{FA} = 10^{-6}$. Note that for Swerling III RCS fluctuations:

$$P_{d \atop SW\,III} = \left[\frac{2}{\left(2+\overline{SNR}\right)^2}\right]\left[2+\left(\left(V_T/\sigma\right)^2/2\right)\overline{SNR}+\overline{SNR}^2/2\right]$$
$$\exp\left[-\left(V_T/\sigma\right)^2/2/\left(1+\overline{SNR}/2\right)\right]$$

Compare these results with the Swerling I calculations for the same SNRs and P_{FA}.

4. Consider a radar with a peak power of 100 kW, 45 dB transmit antenna gain, a 10 square meters antenna aperture, a receiver with system noise (kTB) of –143 dBm, transmit and receive losses of 3 dB each, with a 0 dBsm target at a slant range of 500 km. Using the received power at the radar aperture from the relationship below along with:

$$P_{rec} = \frac{P_t G_t \sigma A_r}{\left(4\pi R^2\right)^2 L_t L_r}$$

and

$$SCR = \left(\frac{A_r}{A_r'}\right)^2 \left(\frac{R_c^3}{R^4}\right) \left(\frac{\sigma}{\sigma^\circ \left(\frac{c\,\tau}{2}\right) \tan \phi \; \theta_{AZ}}\right)$$

the expression for signal-to-clutter ratio for surface clutter, calculate the SNR and SCR for values of σ° of 10^{-M} where $M = 4$, 5, and 6, a clutter range of 20 km, a 10 microsecond pulse length, a depression angle (ϕ) of 6 degrees (coincident with a –20 dB antenna sidelobe), and a 4-degree 3 dB beamwidth in azimuth.

5. For clutter with 3 dB spectral widths of 5, 10, and 15 Hz, calculate the clutter attenuation achieved by two- and three-pulse MTI cancellers with a PRF of 750 Hz, where:

$$I = CA$$

$$I_2 \approx \left[2\left(\pi\sigma_f T\right)^2\right]^{-1} = \left(\frac{PRF}{\sigma_f}\right)^2 \Big/ 19.75 \qquad I_3 \approx \left(\frac{PRF}{\sigma_f}\right)^4 \Big/ 780$$

and where:

σ_f = 1-sigma spectral width of clutter (Hz)
PRF = pulse-repetition frequency waveform (Hz)
T = pulse repetition interval = $1/PRF$ (sec).

6. For the SCRs calculated in problem 1, apply the three-pulse MTI canceller assuming a clutter spectral width of 5 and 10 Hz and a PRF of 750 Hz to calculate the improved SCRs.

Next, calculate the detection probabilities assuming a P_{FA} of 10^{-6} and a Swerling I target model for the resulting signal-to-clutter-plus-noise ratios after using the MTI canceller (use the SNR calculated for problem 1).

Compare these results to the Pd in the clear (i.e., noise-only).

Note the relationship:

$$\frac{S}{C+N} = \frac{1}{\dfrac{1}{SCR} + \dfrac{1}{SNR}} = \frac{SNR \; SCR}{SNR + SCR}$$

Comment on the results. Which cases are clutter-limited? Noise-limited?

7. If each return from a three-pulse transmission follows a Swerling I target model and the desired probability of false alarm is 10^{-6}, calculate the probabilities of detection if coherent integration and 1-out-of-3 detection approaches are employed for SNRs of 5 and 15 dB. Comment on the results.

3

Waveforms, Matched Filtering, and Radar Signal Processing

3.1 INTRODUCTION

This chapter describes the waveforms employed by radars, introduces the concept of the optimal matched filter processor and discusses some common signal processing implementations used in phased-array radars. Topics covered include:

- Representations of complex-valued waveforms
- Fourier transforms of radar signals and their properties
- The matched filter:
  - Simple pulsed continuous wave (CW)
  - Linear frequency modulation (LFM)
  - Phase-code modulation
  - Ambiguity function and diagrams
- Signal processing implementations:
  - All-range digital pulse compression
  - Spectrum analysis or stretch processing of LFM

   – Phase code processing
   – Constant fals alarm rate (CFAR) detection processing
   – Monopulse processing.

There are many excellent references for these topics in the literature. These include references [1], [2], [10], and [12] for general background, while references [5–9], [11], [13], and [15] specifically cover signal theory and signal processing.

3.2 COMPLEX-WAVEFORM REPRESENTATIONS

A waveform can be represented mathematically as:

$$\ell(t) \; = \; r(t)\cos\left[\,2\pi f_0 t \,+\, \theta(t)\right], \tag{3.1}$$

where $r(t)$, f_0, and $q(t)$ are the envelope modulation, operating or center frequency, and the phase modulation, respectively [6, 8]. Using the exponential representation for the cosine yields:

$$\ell(t) \; = \; r(t)\left\{\exp\!\left(j\left[2\pi f_0 t + \theta(t)\right]\right) \,+\, \exp\!\left(-j\left[2\pi f_0 t + \theta(t)\right]\right)\right\}/2, \tag{3.2}$$

where $\ell(t)$, like in the first representation, is still "real valued."

A complex representation of can be stated as:

$$v(t) \; = \; r(t)\exp\left\{j\left[\,2\pi \dot{f_o} + \theta(t)\,\right]\right\}. \tag{3.3}$$

Note that:

$$s(t) \; = \; v(t)\exp\left[-j\left(2\pi f_o t\right)\right] \; = \; r(t)\exp\left[\,j\theta(t)\right]. \tag{3.4}$$

$s(t)$ is termed the "complex envelope" of the original waveform, $\ell(t)$. $s(t)$ is an approximation to $\ell(t)$ for narrowband signals but is close enough for most radar analysis.

The true complex waveform envelope is defined using Hilbert transforms. Also note that:

$$\ell(t) = \frac{1}{2} \text{Re} \left\{ r(t) \exp\left\{ j\left[2\pi f_o t + \theta(t) \right] \right\} \right\}. \qquad (3.5)$$

The complex notation in equation (3.3) through (3.5) and the waveform's complex envelope are useful when analyzing signal processing effects; in other words:

• Use the complex envelope, s(t), for calculations

• Convert back to real-valued waveforms, such as ℓ(t), after otherwise "messy" mathematics are completed.

3.3 FOURIER TRANSFORMS

Fourier transforms and transform theory are useful for manipulating waveforms and signal processing. The Fourier transform is defined for finite energy signals (i.e., for signal x(t), $\int |x(t)|^2 dx$ must be finite) [8]. Note that this is different from the Fourier series, where signals are of infinite energy but are of finite power.

The Fourier transform is defined as:

$$X(f) = \int_{-\infty}^{\infty} x(t) \exp\left(-j2\pi f t\right) dt, \qquad (3.6)$$

with the inverse Fourier transform given by:

$$x(t) = \int_{-\infty}^{\infty} x(f) \exp\left(j2\pi f t\right) df. \qquad (3.7)$$

For example, if x(t) is defined as:

$$x(t) = \begin{cases} 1, & |t| \leq \tau/2 \\ 0, & |t| > \tau/2 \end{cases} \qquad (3.8)$$

then its Fourier transform is given by:

$$X(f) = \int_{-\tau/2}^{\tau/2} \exp(-j2\pi f t) dt = \tau \sin(f\pi\tau)/(f\pi\tau) \equiv \sin c(f\tau). \quad (3.9)$$

Some important and useful Fourier transform properties for time or frequency shifting are:

$$\begin{aligned}
x(t) &\Leftrightarrow X(f) \quad \text{where } \Leftrightarrow \text{denotes Fourier Transform} \\
x(t-t_o) &\Leftrightarrow \exp(-j2\pi f t_o) X(f) \\
X(f-f_o) &\Leftrightarrow \exp(j2\pi f_o t) x(t).
\end{aligned} \quad (3.10)$$

The Fourier transform properties associated with auto correlation (or cross-correlation) are:

or:
$$\begin{aligned}
R_x(\tau) &= \int_{-\infty}^{\infty} x(t) \, x*(t-\tau) \, dt, \\
R_x(\tau) &= \int_{-\infty}^{\infty} |X(f)|^2 \exp[j2\pi f\tau] \, df
\end{aligned} \quad (3.11)$$

Similarly, the transform properties of the often-used Dirac delta function, $\delta(t)$, are given by:

$$\begin{aligned}
f(0) &= \int_{-\infty}^{\infty} f(t) \, \delta(t) \, dt \quad &\Rightarrow \quad I(f) &= \int_{-\infty}^{\infty} \delta(t) \exp[-j2\pi t] \, dt \\
f(t_0) &= \int_{-\infty}^{\infty} f(t) \, \delta(t-t_o) \, dt \quad &\Rightarrow \quad \delta(t) &= \int_{-\infty}^{\infty} \exp[j2\pi f t] \, df
\end{aligned} \quad (3.12)$$

3.4 MATCHING FILTERING

Consider the output of a filter with impulse response $h(t)$ and input $s(t)$:

$$\ell_0(t) = \int_{-\infty}^{\infty} s(t) \, h(\tau-t) \, dt. \quad (3.13)$$

Using the inverse Fourier transform, equation (3.13) can be expressed as:

$$\ell_0(t) = \int_{-\infty}^{\infty} H(f)\, S(f)\, \exp\!\left(j\,2\pi\, f\, t\right) df \, . \tag{3.14}$$

The optimum filter with respect to maximizing the output signal-to-noise ratio (SNR) in white noise is termed the "matched filter" to $s(t)$, and denoted $h_{MF}(t)$ or in the frequency domain representation as $H_{MF}(f)$. The matched filter impulse response is given by [6]:

$$h_{MF}(t) = g^*\, s^*(-t) \Leftrightarrow H_{MF}(f) = g\, S^*(f) \, , \tag{3.15}$$

where g is a complex constant.

The output of the matched filter to an input of $s(t)$ is given by:

$$\ell_{MF}(t) = \int_{-\infty}^{\infty} h(x)\, s(t-x)\, dx = \int_{-\infty}^{\infty} h(t-y)\, s(y)\, dy = \int_{-\infty}^{\infty} s^*(t'-t)\, s(t')\, dt' \, . \tag{3.16}$$

or:

$$\ell_{MF}(t) = \int_{-\infty}^{\infty} s(t)\, s^*(t-\tau)\, dt \, . \tag{3.17}$$

Equation (3.17) is the cross-correlation of $s(t)$ and $s^*(t)$. If the output noise power, P_{noise}, is defined as:

$$P_{noise} = \left(N_0\big/2\right) \int_{-\infty}^{\infty} \left| H_{MF}(f) \right|^2 df \, , \tag{3.18}$$

then the SNR at the output of the matched filter is given by:

$$SNR_{MAX} = \int_{-\infty}^{\infty} \left| S(f) \right|^2 df \, \big/ \left(N_0\big/2\right). \tag{3.19}$$

Since the output of the matched filter is:

$$\ell_{MF}(t) = \int_{-\infty}^{\infty} s(t)\, s^*(t-\tau)\, dt \, , \tag{3.20}$$

it can be represented equivalently as:

$$\ell_{MF}(t) = \int_{-\infty}^{\infty} S^*(f) S(f) \exp(j2\pi f t) df .$$ (3.21)

A conceptual block diagram of equation (3.21) is shown in Figure 3.1. This is a frequency-domain implementation of the matched filter and is equivalent to a time-domain correlator.

Consider a rectangular pulse:

$$s(t) = rect\left(\frac{t}{\tau}\right) \quad \Rightarrow$$

The matched filter output from equation (3.20) is given by:

$$\ell_{MF}(t) = \int_{-\infty}^{\infty} s(t)s^*(t-\tau)dt = \int_{-\infty}^{\infty} rect\left(\frac{t}{\tau}\right) rect^*\left(\frac{t-\tau}{\tau}\right) dt$$

$$= F^{-1}\left\{ F\left\{ rect\left(\frac{t}{\tau}\right)\right\} F\left\{ rect\left(\frac{t-\tau}{\tau}\right)\right\}\right\} = F^{-1}\left\{ \tau^2 \sin c^2(f\tau)\right\}.$$

$$= \tau^2 triang\left(\frac{\tau}{2\tau}\right) \quad \Rightarrow$$

(3.22)

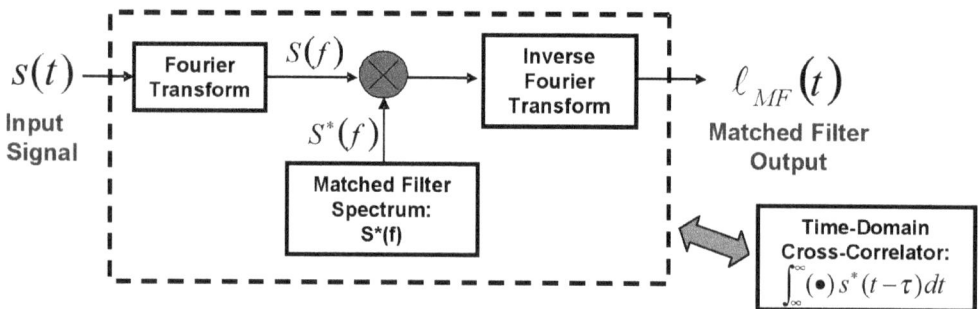

Figure 3.1 Conceptual Block Diagram of Matched Filter Processor

If the signal has voltage spectrum:

$$E_0(f) = \left[\sin(\pi fB)/(\pi fB)\right]^2 \overset{F}{\Leftrightarrow} rect[Bt] * rect[Bt]$$

$$\ell_o(t) = triang(Bt) \Rightarrow$$

(3.23)

A unity-magnitude linear frequency modulation pulse has a complex enve-lope given by:

$$s_{LFM}(t) = \exp\left(j\pi k t^2\right) ; \quad 0 \le t \le \tau ,$$ (3.24)

where the instantaneous frequency is defined as:

$$f(t) = \left(\frac{1}{2\pi}\right)\frac{d\phi}{dt} = kt ,$$ (3.25)

and where its maximum is given by:

$$\Delta = k\tau .$$ (3.26)

The LFM matched filter output can be calculated as:

$$\ell_{MF}(t) = triang\left(\frac{t}{\tau}\right)\left[\sin \pi \Delta t\left(1 - |t|/\tau\right)\right] \Big/ \left[\pi \Delta t\left(1 - |t|/\tau\right)\right] ,$$ (3.27)

which has an approximate "compressed" pulse width of $1/\Delta$ and where near the main lobe of the output the time response is approximately $\sin x/x$. Figure 3.2 il-lustrates the frequency spectrum of an LFM waveform. The complex-conjugate of this spectrum is effectively the desired matched filter frequency response.

The LFM waveform spectrum is given by:

$$S(f) = \int_{-\infty}^{\infty} \exp(j\pi k t^2) \exp(-j\pi f t) \, dt ,$$ (3.28)

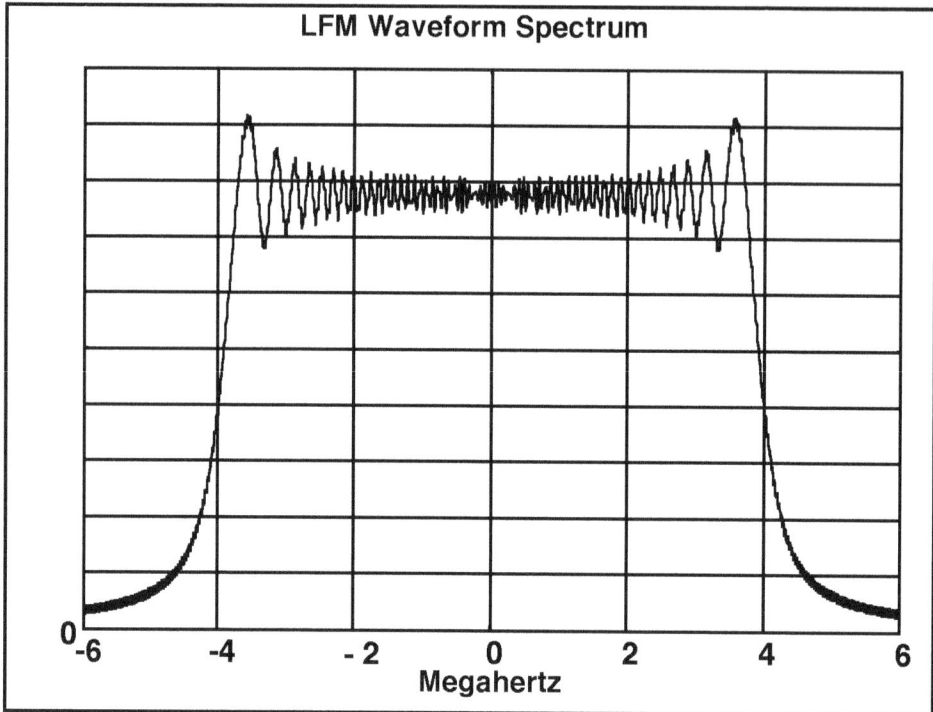

Figure 3.2   LFM Frequency Spectrum (Courtesy of D. P. Harty)

where:

$$\left| S(f) \right| \approx rect\left(\frac{f}{\Delta} \right),$$ (3.29)

with phase:

$$\phi(f) \approx \left(2\pi f^2 \right) / 4\pi k,$$ (3.30)

so that:

$$S(f) \approx rect\left(\frac{f}{\Delta} \right) \exp\left[j\phi(f) \right].$$ (3.31)

Hence, the matched filter to a LFM waveform for large time-bandwidth products is approximately given by:

$$H(f) = rect\left(\frac{f}{\Delta}\right)\exp\left[-j\phi(f)\right] \qquad (3.32)$$

and:

$$E_{OUT}(f) \approx rect\left(\frac{f}{\Delta}\right) \overset{F}{\Leftrightarrow} \ell_{OUT}(t) \approx c\,\frac{\sin\left(\pi\,\Delta t\right)}{\pi\,\Delta t}. \qquad (3.33)$$

3.5 WAVEFORM AMBIGUITY DIAGRAM

The ambiguity diagram is a three-dimensional plot that shows the results of convolving frequency frequency-shifted signals with a fixed reference signal, that is, with the matched filter impulse response. The Doppler frequency-shifted signals represent waveform echoes from moving objects. ([6, 12, 16] for more detailed coverage of this topic.)

An ambiguity diagram depicts two important properties of a waveform: information about a waveform's inherent capability to resolve targets in range and Doppler. The shape of the ambiguity diagram indicates how well suited a waveform is to determine the range and range-rate of an object. The horizontal axes of the diagram are labeled as range (or time delay) and Doppler frequency, and the vertical axis represents magnitude in decibels (dB).

The ambiguity diagram for an LFM pulse is shown in Figure 3.3. The figure illustrates the inherent range-Doppler coupling of the LFM waveform.

3.6 THE FAST FOURIER TRANSFORM (FFT)

The fast Fourier transform (FFT) is an efficient implementation of the discrete Fourier transform commonly used for sampled data. References [7–9] are excellent sources for this material. This section introduces the concept. The FFT is defined as:

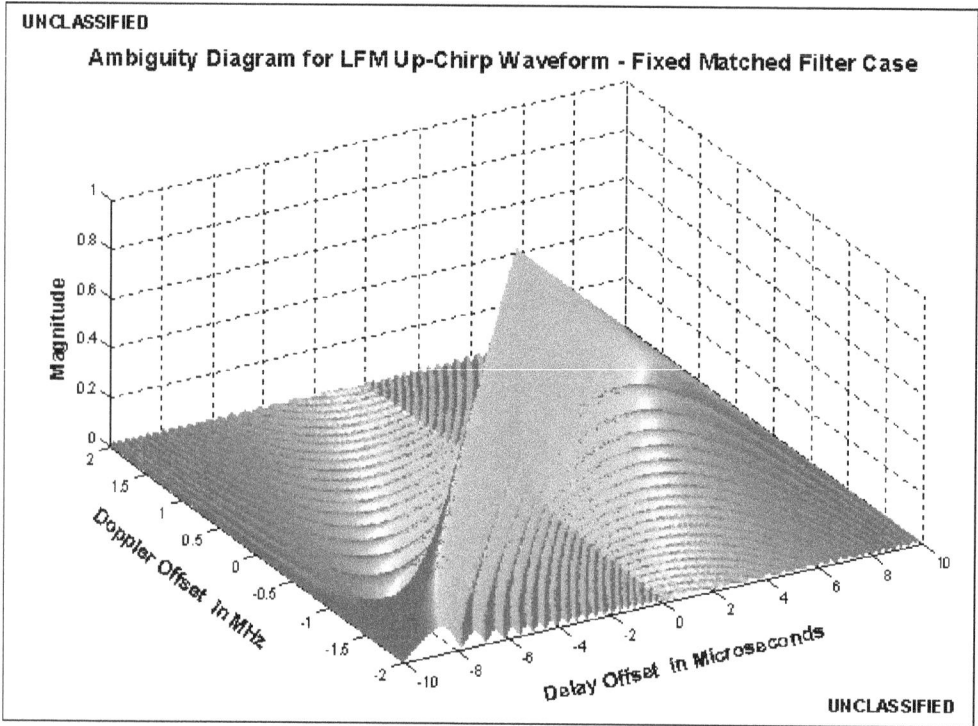

UNCLASSIFIED

Ambiguity Diagram for LFM Up-Chirp Waveform - Fixed Matched Filter Case

UNCLASSIFIED

Figure 3.3   Ambiguity Diagram for a LFM Waveform (Courtesy of D. P. Harty)

$$A(k) = \sum_{n=0}^{N-1} \alpha_n e^{-2\pi j n k / N}. \tag{3.34}$$

The "Cooley-Tukey" FFT algorithm first rearranges the input elements in bit-reversed order, then builds the output transform (decimation in time). Efficiency is achieved by breaking a transform of length N into two transforms of length $N/2$ using the identity where i is used in place of the j used in equation (3.34):

$$
\begin{aligned}
\sum_{n=0}^{N-1} \alpha_n e^{-2\pi i n k/N} &= \sum_{n=0}^{N/2-1} \alpha_{2n} e^{-2\pi i (2n) k/N} + \sum_{n=0}^{N/2-1} \alpha_{2n} \\
&= \sum_{n=0}^{N/2-1} \alpha_n^{even} e^{-2\pi i n k/(N/2)} + e^{-2\pi i k/N} \sum_{n=0}^{N/2-1} \alpha_n^{odd} e^{-2\pi}
\end{aligned}
\tag{3.35}
$$

3.7   DIGITAL IMPLEMENTATION OF MATCHED FILTERS

Figure 3.1 can be implemented for sampled-data in discrete-time using FFTs [7–9]. This is illustrated in Figure 3.4.

Note that Figure 3.4 is a digital implementation of an all-range LFM pulse compression approach. When bandwidths are too large to allow sampling at or above the Nyquist rate, this approach cannot be employed. A technique referred to as "stretch" processing of "spectrum analysis" can be used in these cases, reference [14]. This is depicted in Figure 3.5.

There is a tradeoff between achievable analog-to-digital converter (A/D) sampling rates (and number of bits) and the size of the range window (RW in previous diagrams) to be processed, where the RW is the part of range space over which pulses can be compressed or match filtered.

For a very high LFM bandwidth requirement the pulse cannot be sampled fast enough since A/Ds are not available that run at the required Nyquist rate or higher. Therefore, for very wideband waveforms, all-range digital pulse compression (i.e., two-pass FFT processing) cannot be used. This is the reason for use of spectrum analysis-type techniques (or "stretch" processing), including its one-pass FFT processing implementation where lower A/D rates can be used but limited to smaller RWs (see reference [14]).

When the LFM bandwidth is within the range of current achievable A/D parameters (sampling rates, numbers of bits), then all-range processing (i.e., large RWs) is feasible and is usually used. However, larger RWs do require larger FFT sizes and the associated throughput requirements. For these reasons, the two-pass FFT approach is usually used only for:

- Search (e.g., LFM bandwidths ~ 1 MHz or less) when large range extents are required

- Tracking (e.g., LFM bandwidths ~ 5 to 20 MHz) when large extents are necessary.

When the tracking bandwidth is only 10 or 20 MHz, but only a small RW is necessary (like reacquisition or other functions where the range uncertainty is relatively small), then "stretch" might be used even for lower-bandwidth waveforms. Some radars will employ two types of track processing:

{ $s_{in}(k)$ }

Sampled
baseband
linear FM
(LFM)

FFT

{ $S_{in}(f_k)$ }

**Matched Filter
Weights:**
$H_{MF}(f_k) = S^*(f_k)$

{ $H_{MF}(f_k)$ }

FFT^{-1}

{ $s_{out}(k)$ }

Compressed
pulse output,
i.e., matched
filter output

τ

XMIT

RANGE
WINDOW

RW

time

$N_S = f_S \times (\tau + RW)$; Number of samples
$N_{FFT} = 2^M >= N_S$; FFT dimension

where zero-filling is used for the $N_{FFT} - N_S$ samples

<u>Note:</u> Usually used when $RW >> \tau$ and bandwidth
(B) is low enough that A/D rates can sample >=
Nyquist rate, i.e., search or track bandwidths (e.g.,
< 20 MHz or so)

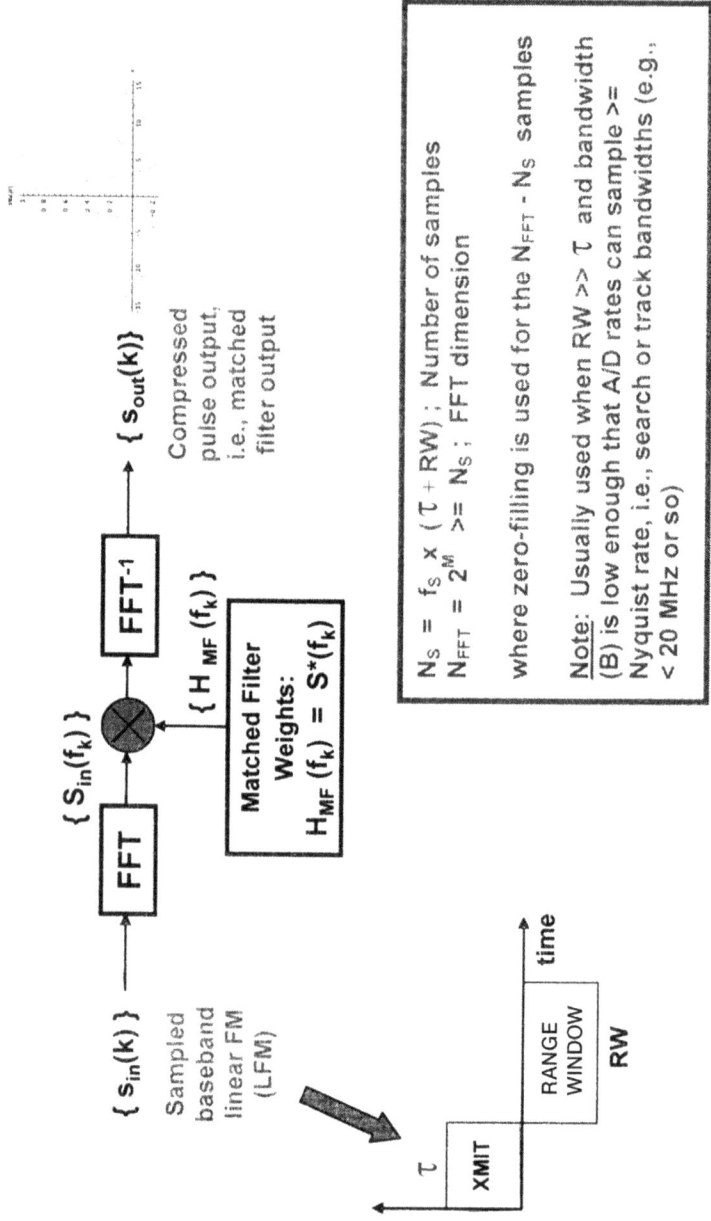

Figure 3.4 LFM Matched Filter Implementation Using FFTs

$\{s_{out}(k)\}$

Compressed pulse output samples, i.e., matched filter output

FFT

$\{s_{de\text{-}ramp}(k)\}$

Sampled de-chirped baseband

I

Q

A/D

A/D

$\cos(w_{IF}\,t)$

$\sin(w_{IF}\,t)$

LPF

"de-chirping" or "de-ramping"

$s_{LFM}(t)$

LFM

$|\,\hat{o} + RW\,|$

RANGE WINDOW

XMIT

time

RW

τ

$N_S = f_S \times (\tau + RW)$; Number of samples

$N_{FFT} = 2^M \geq N_S$; FFT dimension

where zero-filling is used for the N_{FFT} - N_S samples

<u>Note</u>: (i) Usually used when RW << τ and bandwidth (B) is high enough that A/D rates cannot sample >= Nyquist rate, i.e., wideband waveforms (e.g., usually > 50 MHz)

(ii) Matched filtering is distributed across the receiver (i.e., de-ramping) and the signal processor (i.e., spectrum analysis).

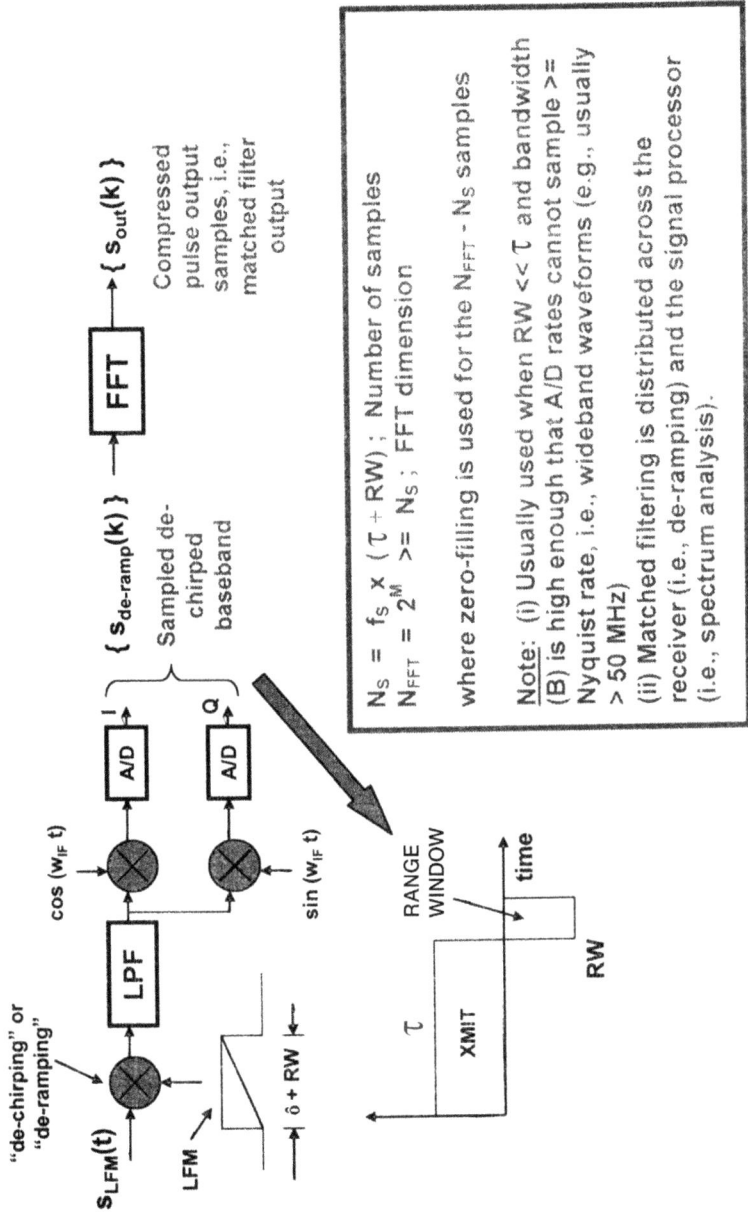

Figure 3.5 Stretch or Spectrum Analysis for LFM Matched Filtering

- A large RW type (larger FFTs required) that uses two-pass (all-range processing)

- A small RW type that uses "stretch" (smaller FFTs required)—this allows many more tracks to be processed.

3.8 PHASE-CODED WAVEFORMS

The LFM waveform achieves a wide bandwidth using analog modulation (or continuous modulation). Phase-coded waveforms obtain a wide bandwidth by using a discrete-phase coding approach. A general phase-coded waveform is of the form shown in Figure 3.6.

The phase, ϕ_i, can be selected from a discrete set of length N, where might be $N = 2^m$ where $m = 1, 2,..., M$. If $N = 2$, the waveform is bi-phase, if $N = 4$, it is quad-phase, etc.

For bi-phase coding:

$$\phi_i = \{0, \pi\}; for\, quad - phase, \phi_i = \left\{0, \frac{\pi}{2}, \pi, \frac{3\pi}{2}\right\}, \qquad (3.36)$$

The same basic matched filter definition is used for phase-coded waveforms, only in discrete form. Hence, integration becomes summation, and the convolution integration is replaced by discrete correlation function:

$$\ell_{OUT}(k) = \sum_{k=1}^{n} s(k)^* s(n-k). \qquad (3.37)$$

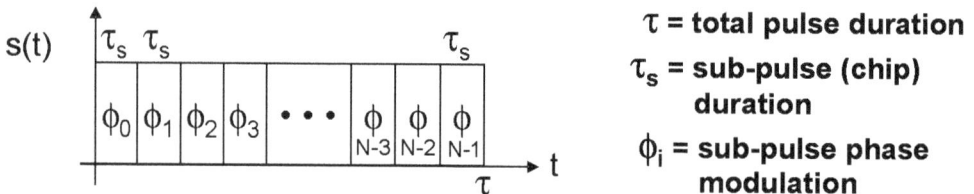

Figure 3.6   Depiction of a Phase-Coded Waveform

The correlator in equation (3.37) can be implemented using a tapped delay line (TDL), where the tap weights are the time-reversed phase modulations as shown in Figure 3.7.

The equivalent bandwidth, B_e, of the phase-coded waveform is defined as:

$$B_e \approx \frac{1}{\tau_s}. \tag{3.38}$$

3.9 WAVEFORM SCHEDULING

Figure 3.8 illustrates some fundamental waveform scheduling concepts, in particular some key definitions. Specifically, the concepts of minimum and maximum range, duty factor, and receive window are defined.

Figure 3.9 illustrates the important concept of timeline occupancy. Occupancy can be considered a radar resource, just as duty factor. In fact, these two quantities are usually those that are "managed" by a radar resource manager and scheduler function.

The example at the bottom of Figure 3.9 represents a radar PRF of 25 pulses/second or, alternatively, 25 beams/second. A radar is defined to be "occupancy-limited" for specified search if:

$$\overbrace{\frac{\psi}{\Omega\, T_{sc}}}^{Re\,quired} > \overbrace{PRF}^{Available}, \tag{3.39}$$

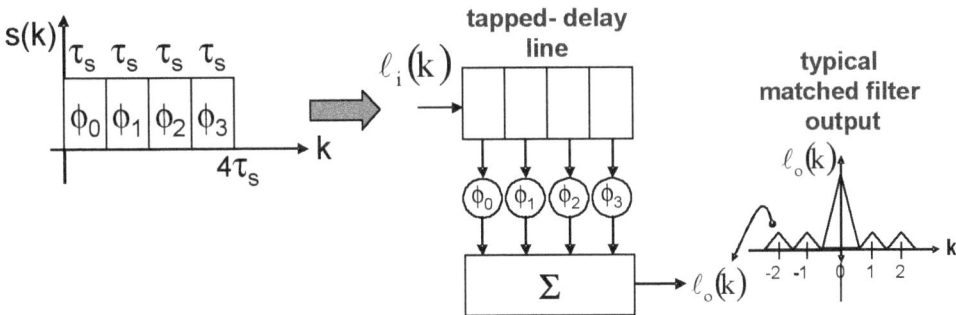

Figure 3.7 Matched Filter for a Phase-Coded Waveform

Tx is the transmit
action

Rx is the receive action
(receive window)

T (resource period)

T_x

0   τ

time

R_x

$R_{min} = c\,\tau/2$ (meters)
(minimum range)

$R_{max} = cT/2 - c\,\tau/2 = c/2(T - \tau)$
(maximum range)

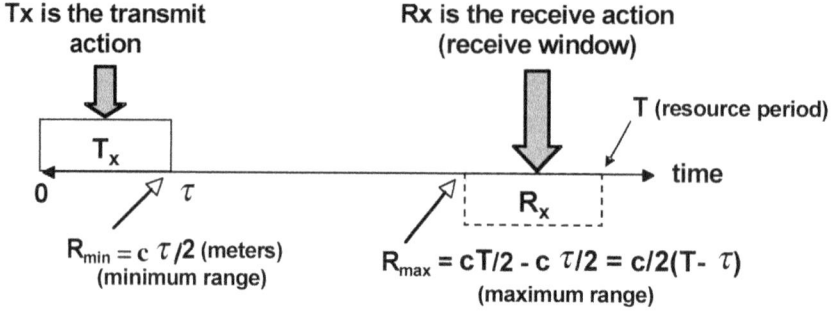

Duty Factor $= \tau / T$; where typical values are in the range 0.1 to 0.25

e.g., for $\tau = 1$ ms and T = 4 ms $\Rightarrow$ Duty Factor $= 0.25$ or 25%

Figure 3.8 Waveform Scheduling Concepts and Definitions

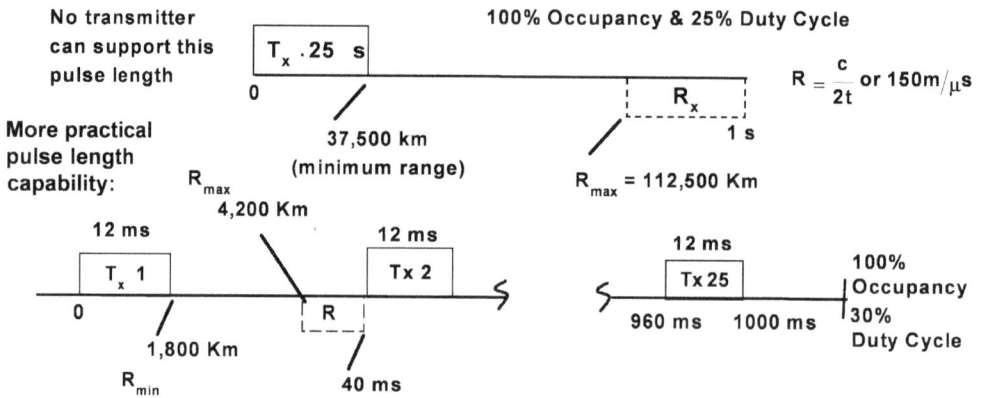

No transmitter
can support this
pulse length

T_x .25 s

100% Occupancy & 25% Duty Cycle

0

$R = \dfrac{c}{2t}$ or $150m/\mu s$

R_x

More practical
pulse length
capability:

37,500 km
(minimum range)

1 s

R_{max}
4,200 Km

$R_{max} = 112,500$ Km

12 ms

T_x 1

0

12 ms

Tx 2

12 ms

Tx 25

100%
Occupancy

R

960 ms   1000 ms

30%
Duty Cycle

1,800 Km

R_{min}

40 ms

Figure 3.9 Illustration Radar Timeline Occupancy Concept

where Ψ, Ω, and T_{SC} are the search volume in radians^2, the radar antenna beam area in radians^2, and the specified scan or search frame time, respectively, and PRF is the pulse repetition frequency.

3.10 WAVEFORMS AND RADAR FUNCTIONS

The waveforms that are typically employed for common radar functions are:

- Search:
  - Un-coded CW
  - Narrow band linear frequency modulation (LFM)
  - 500 kHz to 1 MHz bandwidth (i.e., very narrowband)

- Track initiation/track maintenance:
  - LFM
  - 5 to 20+ MHz bandwidth (i.e., narrowband)

- Target classification:
  - LFM
  - Wide bandwidths.

As can be seen, very narrowband waveforms are usually employed for search, narrowband waveforms are used for tracking, and wideband waveforms are used for classification.

3.11 OTHER RADAR SIGNAL PROCESSING FUNCTIONS

This section describes two signal processing functions in addition to the waveform matched filtering described in previous sections. Constant false alarm rate (CFAR) and monopulse processing are described in the following subsections.

3.11.1 Constant False Alarm Rate Processing

CFAR processing is a signal processor algorithm that sets the detection threshold for declaring target returns [6, 17]. This selection of the detection threshold affects the false alarm performance and detection performance of a radar system. In thermal noise-only environments (no clutter or jamming), the detection threshold can be calculated from equations (3.40) and (3.41), where P_{FA} and $2\sigma^2$ are the probability of false alarm and the thermal noise power, respectively:

$$P_{FA} = \int_{-V_T}^{\infty} \frac{\alpha}{\sigma^2} e^{-\left(\frac{\alpha}{\sigma}\right)^2 / 2} \, d\alpha = e^{-\frac{V_T^2}{2\sigma^2}} \tag{3.40}$$

and:

$$V_T^2 \;=\; -\,2\,\hat{\sigma}^2 \ell n\!\left(P_{FA}\right). \tag{3.41}$$

Figure 3.10 shows the Rayleigh probability density associated with the random thermal noise magnitudes.

However, in most applications, the noise power neither is known precisely nor is constant. This necessitates estimating the localized (in range and angle) noise floor before computing the threshold using measurements of the background noise.

When the environment consists of noise plus homogeneous or nonhomogeneous clutter and/or jamming, the estimates of background interference are needed. The CFAR processor is designed to perform these functions—that is, to estimate the background interference level and "bias" it to achieve the desired P_{FA}.

Figure 3.11 depicts an example of a cell-averaging CFAR, referred to as a CA-CFAR. A general configuration for a CFAR is depicted in Figure 3.12. Received target echoes are amplified (linearly or logarithmically), detected, and then passed through a tapped delay line. Usually "early" and "late" background noise or interference levels are estimated using the "late" and "early" TDL cells used with some combining logic (usually selectable). The "cell under test" (CUT) is compared with the calculated threshold using a "biased" noise/interference estimate to achieve the desired P_{FA}.

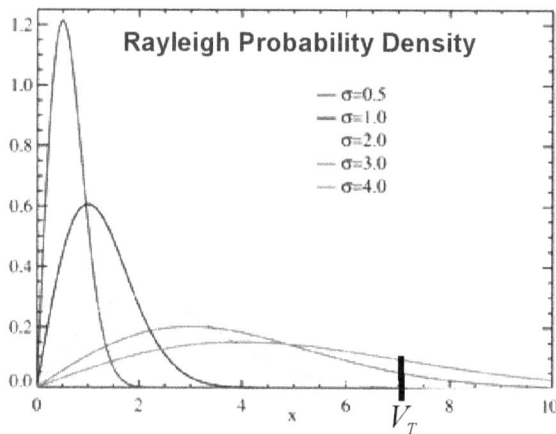

Figure 3.10 Probability Density for Thermal Noise

Radar
Matched
Filter Output
Range Cells
(linear or
logarithmic)

Leading CFAR
Sliding Window

CFAR Guard
Cells

Cell-under-
test (CUT)

Lagging CFAR
Sliding Window

$\oplus$

Average
Estimate*

Leading
Estimate of Noise;
$\hat{N}_{LD}$

$\oplus$

Average
Estimate*

Lagging
Estimate of Noise:
$\hat{N}_{LG}$

Note:
* Average may
censor M largest
cells from result

Threshold
Computation **

$\hat{V}_T$

$$CUT \underset{H_1}{\overset{H_0}{\gtrless}} \hat{V}_T$$

where H_0 = {No target}
H_1 = {Target present}

Note:
** Threshold may employ
Leading, or Lagging,
or greatest of or
Least-of logic

* The basic
flow includes:
* Leading cells
* Guard cells
* Cell-under-test
* Lagging cells

* Use of $\hat{V}_T$ incurs a loss in
detection sensitivity (SNR);
i.e., the CFAR loss.

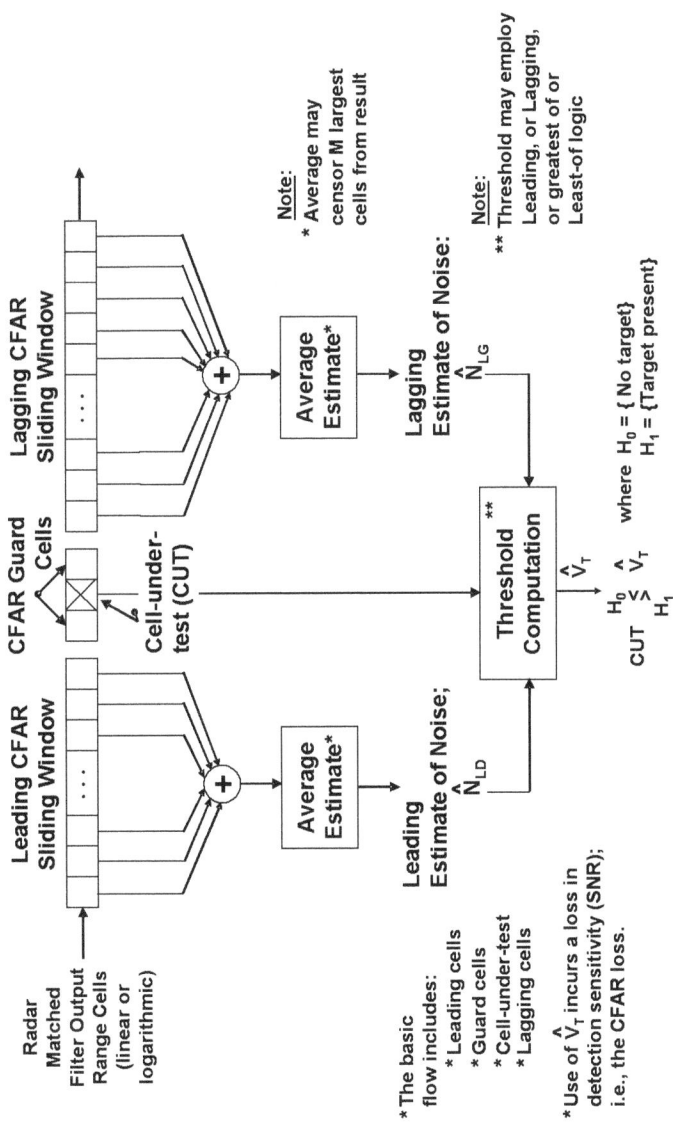

Figure 3.11   Example of a Cell-Averaging CFAR (CA-CFAR)

73

Signal From Antenna/ Receiver

```
Signal From Antenna/Receiver
        │
        ▼
┌──────────────┐   ┌──────────┐   ┌──────────────────┐
│ Linear or    │   │ Envelope │   │ Tapped Delay Line│
│ Logarithmic  │──▶│ Detector │──▶│                  │
│ Amplifier    │   │          │   │                  │
└──────────────┘   └──────────┘   └──────────────────┘
```

"Late" CFAR Cells

"Cell" Under Test (CUT)

V_{-N} V_{-N-1} $\cdots$ V_0 $\cdots$ V_{N-1} V_N

"Early" CFAR Cells

Threshold Setting (Rule-Based)

Estimate Based on Selected Cells

Comparator

CFAR Output

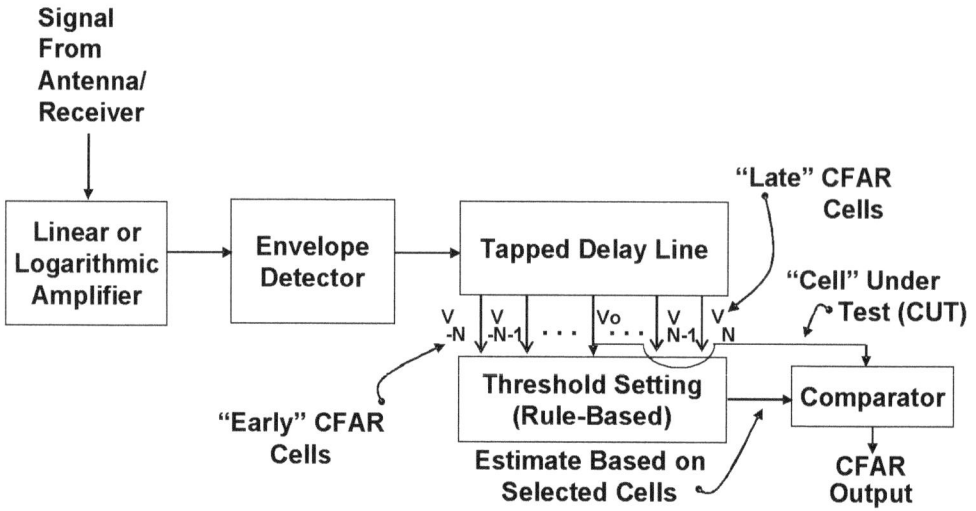

Figure 3.12   Generalized CFAR Processor Block Diagram

The most common form of CFAR is the cell-averaging type (i.e., CA-CFAR). Usually "early" and "late" cells are used to compute early and late "window" noise averages. These may be simple arithmetic averages, or they may seek to exclude some cells from the computed average. Typically excluded cells include:

• Cells on either side of the CUT (e.g., 1 to 3 or so)

• Cells that exhibit large values (spiky interference or noise); often the largest two or three values are excluded.

If M cells remain after "editing," the noise estimate (for linear) processing is given by:

$$\overline{P_r} = \left(\sum_{m=1}^{M} q_m \right) \Big/ M \,. \tag{3.42}$$

Note that for logarithmic processing, the estimate may be computed using:

$$\overline{P}_{r-\log} = \left(\sum_{m=1}^{M} \log_2 q_m \right) \Big/ M \,. \tag{3.43}$$

The logarithm of any base can be used in place of the base-2 logarithm indicated in equation (3.43).

Since the noise (or interference) power estimate uses a finite (M) number of cells, there is a loss in detectability, on average, when using the *linear* CFAR processor for detection. For Swerling I type target fluctuations, the loss is reflected in the P_D:

$$P_D = \left\{ (1+\gamma_a) \Big/ \left[1+\gamma_a + \left(P_{FA}^{-\frac{1}{M}} - 1 \right) \right] \right\}^M ; \quad \text{where} \quad \gamma_a = \frac{\ell n \left(\dfrac{P_{FA}}{P_d} \right)}{\ell n (P_D)} .$$

$$(3.44)$$

This can be compared to the ideal P_D in noise for Swerling I targets:

$$P_D = (P_{FA})^{\frac{1}{1+SNR}} = (P_{FA})^{\frac{1}{1+\gamma_a}} . \qquad (3.45)$$

The signal-to-interference plus noise ratio (SINR) for M-cell estimates is therefore:

$$\gamma_a(M) = \left[(P_D/P_{FA})^{\frac{1}{M}} - 1 \right] \Big/ \left(1 - P_D^{\frac{1}{M}} \right). \qquad (3.46)$$

The difference between $\gamma_a(M)$ and the ideal

$$SNR = \frac{\ell n (P_{FA})}{\ell n (P_D)} - 1$$

is referred to as the (homogeneous environment) CFAR loss, which in decibels is approximately given by:

$$L_{CFAR}(dB) = 10 \log_{10} \left[\gamma_a(M)/\gamma_a(\infty) \right]. \qquad (3.47)$$

In the previous homogeneous environments, it was assumed that the background was statistically "flat" or having a near constant mean with a relatively small variance. However, the interference background can often consist of discrete values (such as other targets), extended clutter, and other discrete returns. In these environments, the false alarm probabilities achieved will vary, sometimes considerably from their desired value. In addition, the P_D may not increase (i.e., reduce the CFAR loss) when larger numbers of cells are used.

Alternative techniques are required to achieve the desired CFAR false alarm performance. However, this performance is obtained at the cost of a greater loss in detectability. Examples of nonhomogeneous backgrounds include: step functions, ramps (up-constant-down) or trapezoidal envelopes.

The CA-CFAR cannot maintain constant false alarm performance during significant changes in background levels (e.g., transitions for step functions, ramps up and down for trapezoidal shapes). The P_D obtained in nonhomogeneous backgrounds, such as "clumps" of clutter, is given by:

$$ P_D = \prod_{m=1}^{M} \left\{ \left[1 + P_m \left(P_{FA}^{-\frac{1}{M}} - 1 \right) \right] \Big/ \left[P_0 (1 + \gamma_a) \right] \right\}^{-1}, \qquad (3.48) $$

where:

P_0 = total interference power of test cell

P_m = total interference of reference cell M

Note that if test and reference cells all have the same powers, this reduces to the previous result:

$$ P_D = \left\{ (1 + \gamma_a) \Big/ \left[1 + \gamma_a + \left(P_{fa}^{-\frac{1}{M}} - 1 \right) \right] \right\}^{M}. \qquad (3.49) $$

The linear CFAR, although reasonably optimum for homogeneous backgrounds is suboptimum for nonhomogeneous backgrounds, due to sharp transitions in background interference levels. This suboptimum estimation performance can be improved by using the logarithmic CFAR. For this type of CA-CFAR, with "ramp" or transition portions of the interference background, the false alarm probability is independent of test cell power and slope of the ramp.

The required SINR necessary to achieve a given P_D and P_{FA} for logarithmic CFAR is greater than that for the linear CA-CFAR in homogeneous background. Hence, geometric mean (or logarithmic CA-CFAR) processors exhibit higher CFAR losses. However, the higher loss is traded off against ensuring the desired P_{FA} performance in nonhomogeneous interference backgrounds.

Although CA-CFAR is optimum in homogeneous noise and jamming, the optimality of geometric mean CFAR is not known. Also, the type of CFAR necessary to achieve "best" performance varies with the type of background present. Alternative CFARs developed to deal with different nonhomogeneous interferences include:

- Greatest-of (GO-CFAR)

- Order Statistics (OS-CFAR)

- Censoring (already discussed)

- Composite (combinations of the above and other forms)

"Greatest-of" uses the larger of the "early" and "late" averages to compute the threshold. This results in a slight increase in CFAR loss when used in homogeneous environments.

Other types of nonhomogeneity are the presence of additional targets in the reference cells. An M-cell CA-CFAR with K additional targets present results in a degraded P_D of:

$$P_D = \left[1 + k_a/\left(1 + \gamma_A\right)\right]^{-(M-K)}\left[1 + k_a\left(1 + \gamma\right)/1 + \gamma_A\right]^{-K}, \qquad (3.50)$$

where k_a is the adaptive threshold multiplier, γ is the SINR of the test cell (CUT), and γ_A is the SNR of the additional targets. The multiple-target-in reference cells condition results in higher CFAR losses. The loss can be reduced by using the geometric mean, censoring (discussed previously), use of the median to form the threshold, or more generally using a multiple of an arbitrary order statistic of the reference cells (OS-CFAR). The CFAR loss for a 32-cell CFAR for $P_{FA} = 10^{-6}$ is given by:

CA-CFAR	0.97 dB
GO-CFAR	1.13 dB
OS-CFAR (75% rank)	1.45 dB
GO-OS-CFAR (75% rank)	1.66 dB
CA-CFAR, 1-cell censor	1.01 dB
CA-CFAR, 2-cell censor	1.06 dB.

$$\text{(3.51)}$$

For slowly varying clutter, a smoothed (recursive) clutter map can be computed for range and angle cells. The recursive estimate is, for the kth cell found using:

$$Q_k(m) = (1 - w) Q_k(m - 1) + w \, q_m(k), \qquad (3.52)$$

where $q_m(k)$ is the kth cell amplitude, w is the smoothing weight, and m is the "scan" number. For this clutter map estimate, P_d is given by:

$$P_D = \pi \sum_{r=0}^{\infty} \left[1 + k_c w (1 - w)^r / (1 + \gamma_a) \right]^{-1}, \qquad (3.53)$$

where k_c is the threshold multiplier for a given P_{FA} (iteratively determined with γ_a set to zero).

The objective of using composite techniques is to reduce the CFAR loss. Effectively, by switching CFAR types based on measured averages compared to a threshold (or other rule), the lowest-loss CFAR that achieves desired P_{FA} performance is employed. An example would be combining a CA-CFAR for noise-only and a censoring CFAR for multiple targets present, or log CA-CFAR for "clumps" of clutter.

3.11.2 Monopulse Processing

Another key technique used by radars is monopulse. Monopulse is used to extract angle measurements target returns [18]. The maximum-likelihood angle estimate is given by:

$$\hat{\theta}_{ML} = \frac{\theta_3}{k_m} \mathrm{Re}\left\{\frac{\Delta}{\Sigma}\right\},$$ (3.54)

where θ_3, k_m, Σ, and Δ are the antenna 3 dB beamwidth, monopulse slope, and antenna sum and difference-channel voltages, respectively, and Re{} denotes the real part of a complex number.

Monopulse accomplishes this function by separating two-dimensional receive antennas into azimuth and elevation "quadrants" and combines them to estimate angular positions. An "error" pattern is formed as $e(\theta) = \Delta(\theta)/\Sigma(\theta)$ as shown in Figure 3.13.

The amplitude and sign of $e(\theta)$ indicates the distance of the target from the antenna pointing angle.

The thermal noise-limited one-sigma angle accuracy associated with the monopulse technique is the familiar Cramer-Rao-type bound:

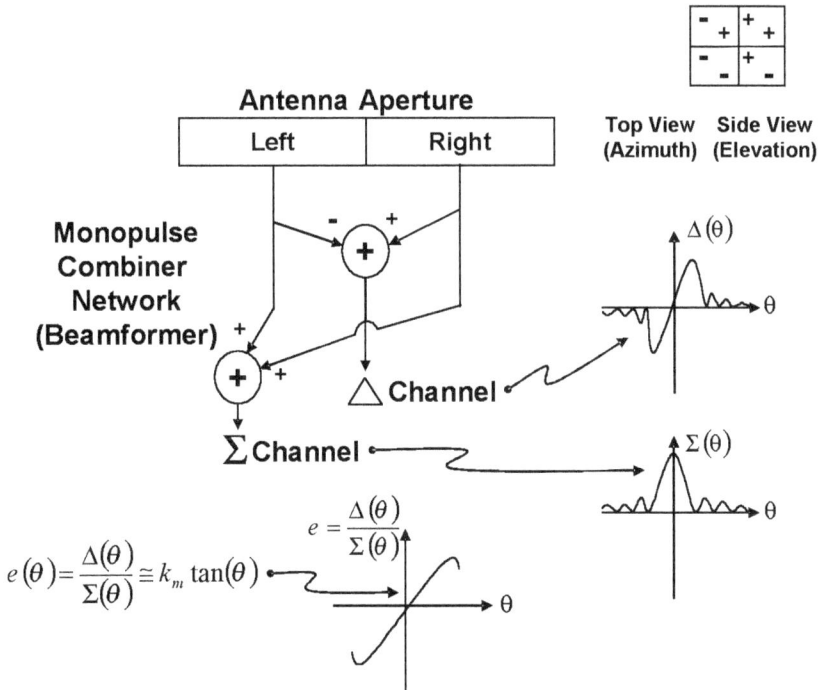

Figure 3.13 Diagram of Basic Monopulse Technique

$$\sigma_\theta = \frac{\theta_3}{k_m} \frac{1}{\sqrt{2\ SNR}} , \tag{3.55}$$

where SNR is the signal-noise-ratio at the matched filter and monopulse comparator output.

The complex-valued monopulse ratio also can provide an indication if there are two or more unresolved targets in a range-cell but within the 3 dB antenna receive beamwidth. A hypothesis test similar to that used for target detection can be employed for this purpose:

$$\gamma_{imaginary} = \mathrm{Im}\left\{\frac{\Delta}{\Sigma}\right\} \begin{array}{c} H_1 \\ > \\ < \\ H_0 \end{array} T_{unresolved} , \tag{3.56}$$

where Im{} denotes the imaginary part of a complex-valued quantity, H_0 and H_i are the hypotheses for resolved and unresolved targets, respectively, and $T_{unresolved}$ is the test threshold. In a similar manner to the target detection problem, the threshold is selected to achieve a specified probability of false "unresolved" indication. Target returns with values of $\gamma_{imaginary}$ that exceed the threshold would not be used for tracking or discrimination, for example, where the corrupted return data could potentially degrade radar performance.

3.12 REFERENCES

[1] D. K. Barton, *Radar System Analysis and Modeling*, Artech House, 2004
[2] E. Brookner, *Aspects of Modern Radars*, Artech House, 1988
[3] S. Haykin, *Adaptive Radar Signal Processing*, Wiley-Interscience, 2006
[4] S. Kay, *Modern Spectral Estimation: Theory and Application*, Prentice-Hall, 1999
[5] N. Levanon, *Radar Principles*, Wiley-Interscience, 1988
[6] R. Nitzberg, *Radar Signal Processing and Adaptive Systems*, 2nd Edition, Artech House, 1999
[7] A. Oppenheim & R. Shafer, *Digital Signal Processing*, Prentice-Hall, 1975
[8] A. Papoulis, *Signal Analysis*, McGraw-Hill, 1977
[9] L. Rabiner, *Theory and Application of Digital Signal Processing*, Prentice-Hall, 1975
[10] M. Skolnik, *Introduction to Radar Systems*, 3rd Edition, McGraw-Hill, 2002
[11] S. Haykin & A. Steinhardt, *Adaptive Radar Detection and Estimation*, Wiley, 1992
[12] D. K. Barton, *Modern Radar System Analysis*, Artech House, 1988

[13] D. R. Wehner, *High Resolution Radar*, Artech House, 1987

[14] W. J. Caputi, "Stretch: A Time Transformation Technique," *IEEE Transactions on Aerospace Electronic Systems*, AES-7, March 1971

[15] J. V. Candy, *Signal Processing—The Modern Approach*, McGraw-Hill, 1988

[16] A. W. Rihaczek, *Principles of High-Resolution Radar*, Mark Resources, 1977

[17] G. Minkler & J. Minkler, *CFAR*, Magellan, 1990

[18] S. M. Sherman, *Monopulse Principles and Techniques*, Artech House, 1984

3.13 PROBLEMS

1. Consider a radar waveform with a triangular-shaped pulse as shown below. Calculate the matched filter output for this waveform. Sketch the matched filter output waveform (hint: start with the analysis from chart 10 and apply the convolution property of Fourier transforms).

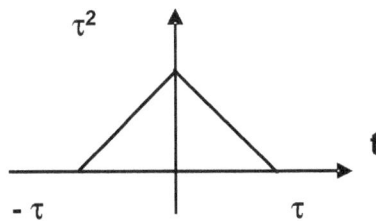

2. Estimate the approximate 3 dB range resolution of the matched filter output in problem 1 if $\tau = 10$ μsec. What is the approximate bandwidth of the matched filter output?

3. If a linear FM (or "chirp") waveform is used to achieve the above range resolution, what bandwidth is required? If no time sidelobe weighting is applied during matched filtering, compare with the sidelobes of the waveform from problem 1. Is the waveform from problem 1 realizable? If a triangular taper is applied to the LFM waveform, what is the amplitude of the first sidelobe? What penalties are incurred when using the triangular taper?

4. Consider a radar that employs a 1 millisecond (τ) 1 MHz LFM for search. All-range digital pulse compression is to be used for matched filtering to cover a 10 km range window (RW). Since some degree of interpolation is desired, a complex sampling rate of 1.2 MHz is to be used. Calculate the size of the fast Fourier transform (FFT) required for the signal processing (assuming

that zero-filling will be used to pad out the FFT to a power of 2, i.e., $2^M \geq N_s$). For calculation purposes, note that for pulse compression:

$$N_s = \left(f_{samp}\right)\left(\tau_p + RW\right) \qquad N_{FFT} = (2)^M .$$

5. For tracking purposes, a 1 millisecond 20 MHz LFM is used. A "stretch" or "spectrum analysis" approach is employed for matched filtering over the 1 km range window. Calculate the required complex sampling rate necessary to implement this technique and the required FFT size. For calculation purposes, note that for the "stretch" process:

$$B_s = \left(\frac{B}{\tau_p}\right)(RW) .$$

6. Consider a CFAR processor that uses a 16 range cells to estimate the background noise-plus-interference level. For a required probability of false alarm of 10^{-6}, calculate the resulting probability of detection when probability of detection using the ideal noise threshold is 0.9 for a Swerling I target.

$$P_D = \left\{(1+\gamma_a)\Big/\left[1+\gamma_a+\left(P_{FA}^{-\frac{1}{M}}-1\right)\right]\right\}^M ; \quad where \ \gamma_a = \frac{\ln\left(\dfrac{P_{FA}}{P_D}\right)}{\ln(P_D)} .$$

7. Estimate the sensitivity loss (i.e., loss in equivalent SNR) that results when using the 16-cell CFAR. Note the relationship for a Swerling I target using an ideal noise threshold is:

$$P_D = \left(P_{FA}\right)^{\frac{1}{1+SNR}} .$$

8. Consider the CFAR processor and parameters used in problem 1 in a non-homogeneous environment. Assuming that the background level is identical in each range cell, calculate the resulting probability of detection and compare with the result in problem 6.

$$P_D = \left\{ \left(1 + \gamma_a \right) \middle/ \left[1 + \gamma_a + \left(P_{FA}^{-\frac{1}{M}} - 1 \right) \right] \right\}^M .$$

9. Consider the use of a 32-cell CFAR processor to detect a Swerling I target in a homogeneous noise background. If the SNR is 15 dB and the desired probability of false alarm of 10^{-6}, calculate the resulting probabilities of detection when cell-averaging (CA) CFARs with and without 2-cell censoring, a greatest-of (GO) CFAR, and a GO-ordered statistics (OS) CFAR are used. Compare this to the case of an ideal noise threshold as calculated in problem 7.

4

Search and Acquisition Functions

4.1 INTRODUCTION

This chapter treats the various types of radar searches, such as volume search, horizon fence search, cued search, and sector search, the different types of waveforms used, and the acquisition function used as a prerequisite for track initiation (TI). Topics include:

- Types of searches:
  - Volume
  - Horizon fence and sector search
  - Cued
  - Multibeam
- Search design
- Search waveforms and processing:
  - Surveillance
  - Verification
- Acquisition waveforms and processing.

As in previous chapters, there is a wealth of information on search and acqui-sition in the radar literature. Relevant references are cited where appropriate throughout the chapter. These sources provide the theory necessary to design and analyze radars that perform these functions.

4.2   TYPES OF SEARCHES

In this section, a number of commonly employed searches are described for use in either autonomous (e.g., large volume searches) or nonautonomous applica-tions (e.g., handover or cued searches).

4.2.1   Volume Search

Volume search was briefly introduced in Chapter 1, and two forms of the radar range equation (RRE) were derived and discussed in Section 1.6.2. This and sub-sequent sections will build upon the material in Chapter 1 and focuses on some specific application-oriented design guidance for volume search.

Figure 4.1 is identical to that included in Chapter 1 and illustrates a typical three-dimensional (3-D) search volume.

Volume searches are defined by the parameters:

- Search coverage volume, Ψ (degrees^2 or radians^2)

- Number of antenna beams, N_b equivalent to coverage for a given aperture size and operating frequency

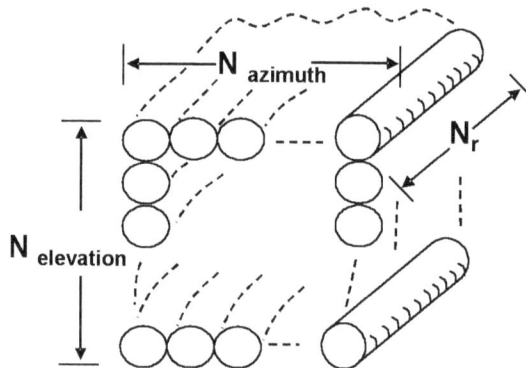

Figure 4.1   Volume Search Beam Raster for Phased-Array Radar

- Search range extent, ΔR (km)
- Frame or scan time, T_{SC} (s)
- Search waveform bandwidth, B_s (MHz)
- Allowable false alarm rate, η_{FA} (s^{-1})
- Desired probability of detection, P_D.

From the number of antenna beams needed to provide the desired angular coverage, range extent, search waveform bandwidth, and false alarm rate, the allowed probability of false alarm can be calculated as:

$$P_{FA} = \left(\frac{2 B_s}{c} \right) \Delta R \, N_b \, \eta_{FA} \, T \; . \tag{4.1}$$

Given the P_D, P_{FA}, and the type of target fluctuation model (e.g., Swerling I, log-normal, etc.), the necessary signal-to-noise ratio (SNR) for search can be calculated. Once the desired SNR is known, then the required power-aperture product (PA) is given by the volume search form of the radar range equation (RRE), introduced in Chapter 1:

$$SNR = \frac{\sigma \, T_{SC}}{\left(4 \pi \right) k T_s \, R^4 \, \Psi \, L_t \, L_r} \, P_{AVE} \, A_r \; . \tag{4.2}$$

Solving equation (4.2) for the required power-aperture product:

$$P_{AVE} \, A_r = \frac{\left(4 \pi \right) k T_s \, R^4 \, \Psi \, L_t \, L_r \, SNR}{\sigma \, T_{SC}} \; , \tag{4.3}$$

where Ψ is the area to be searched in radians^2 and T_{SC} is the scan or "frame" time for the search. Equation (4.3) is the familiar relationship that shows that the inherent capability of a radar to perform volume search is operating frequency-agnostic. Table 4.1 provides an example that calculates the power-product needed to perform a specified volume search (60 degrees in azimuth by 10 degrees in elevation) using a logarithmic form of equation (4.3). The result from

Table 4.1   Volume Search Design Example

Radar parameter	+	−
4π	10.99	
kT_s		200.81
R^4 (500 km)	227.95	
Ψ (60 degrees by 10 degrees)		7.38
L_t (3dB)	3.00	
L_r (3 dB)	3.00	
SNR (12 dB per pulse)	12.00	
σ (1 meter^2)	0.00	
T_{SC} (5 seconds)		6.99
Totals	256.94	215.18

Table 4.1 is a required power-aperture product of 44.52 to 41.76 dBW-m^2. So for a 10 m^2 antenna aperture, the required average power is 34.52 to 31.76 dBW or equivalently, 2.83 to 1.50 kW. For a 25% duty factor, this corresponds to a peak transmitter power of 11.33 to 6.0 kW.

The waveforms for volume search are typically narrowband (e.g., 500 kHz to 1 MHz). However, since they do usually necessitate the processing of large range extents (or range windows), this impacts the needed signal processing. Many search radars employ linear frequency modulation (LFM) waveforms to perform volume search. Due to the aforementioned large range extents that are often required, the matched filters used are usually all-range digital pulse compression, as described in Chapter 3.

The typical search approach employs verification waveforms to corroborate the search detections. These are usually narrowband waveforms (e.g., the same as the search waveform). Wise allocation of radar resources between search and verification functions can reduce the radar's energy expenditure for search and acquisition. This is achieved through use of a lower P_{FA} for the verify waveform sequence than used for search.

4.2.2   Horizon Search Fence

This type of search is used by radars that perform ballistic missile early warning (EW) or missile defense (MD) functions. As described in Section 6.2 of Chapter 1, it is useful for detection and acquisition of ballistic missiles within the detection range of the radar. The premise is that targets with positive elevation rates will "break" the horizon fence and result in detection opportunities. As can be seen, the use of a single row of antenna beams at or slightly above the local horizon can result in a minimum-energy solution for missile target detection. Figure 4.2 illustrates a typical horizon search fence used for these applications.

Cumulative detection probability is the fundamental theory exploited by these searches; that is, multiple detection opportunities result in a high aggregate probability of target detection. Again, some form of verification process is used prior to track initiation to mitigate the wastage of radar resources attempting to initiate track on false alarms. However, this is fundamentally identical to the method used for volume searches.

As identified in Chapter 1, the radar range equation for horizon search is given by:

$$SNR = \frac{\sigma}{\left(2\sqrt{\pi}\right) kT_s \, R^3 \, \Psi \, N \, v_T \, L_t \, L_r} \frac{P_{AVE} \, A_r}{\sqrt{G_r}} , \qquad (4.4)$$

where G_r is the receive antenna gain, v_T is the vertical rate of the target in meters/second, and N is the number of looks required for detection. Solving equation (4.4) for the power-aperture product yields:

$$P_{AVE} \, A_r = \frac{\left(2\sqrt{\pi}\right) kT_s \, R^3 \, \Psi \, N \, v_T \, L_t \, L_r \, \sqrt{G_r} \, SNR}{\sigma} . \qquad (4.5)$$

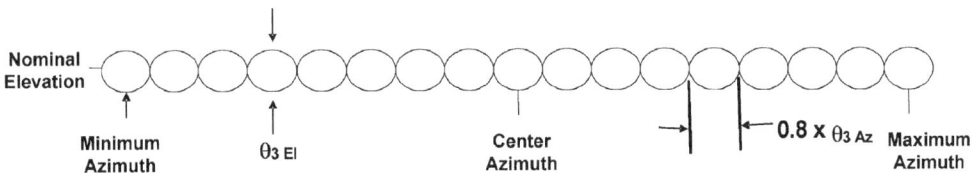

Figure 4.2   Typical Horizon Search Fence

As can be seen, since the receive antenna gain is a function of the receive aperture, equation (4.5) must be solved iteratively. Alternatively, it can be re-expressed as:

$$P_{AVE} A_r = \frac{0.886 \left(4\pi\right) kT_s \, R^3 W N v_T \, L_t \, L_r \, SNR}{\sigma} \, , \qquad (4.6)$$

where W is the total fence azimuth extent in radians.

Table 4.2 provides an example that calculates the power-product needed to perform a ±30 degree horizon fence search using a logarithmic form of equation (4.6).

Table 4.2   Horizon Search Design Example

Radar parameter	+	−
0.886 (4π)	10.46	
kT_s		200.81
R^3 (500 km)	170.96	
W (60 degrees)	0.20	
L_t (3 dB)	3.00	
L_r (3 dB)	3.00	
SNR (12 dB per pulse)	12.00	
σ (1 meter²)	0.00	
N (3 looks)	4.77	
v_T (2200 meters/second)	33.42	
Totals	237.81	200.81

The result from Table 4.2 is a required power-aperture product of 37.0 dBW-m². So, as before, for a 10 m² antenna aperture, the required average power is 27.0 dBW or equivalently, about 503.65 W. For a 25% duty factor, this corresponds to a peak transmitter power of 2.0 kW. Note the lower power-aperture requirement for the horizon search fence as compared with that for the previous volume search. This is due to the horizon fence searching only a single row of antenna beams as opposed to 10 degrees of elevation coverage provided in the volume search example.

A horizon search fence typically uses narrowband waveforms for initial detection as well as for verification purposes. These are often LFM waveforms with bandwidths in the range of several hundred kHz and slightly higher. In general, due to the large range extents to be searched, all-range digital pulse compression processing is employed for these waveforms. Like volume searches, horizon search fences typically use noise thresholds for detection purposes as opposed to constant false alarm rate (CFAR) to minimize the loss in sensitivity (i.e., SNR) associated with the latter approach.

Sector searches are similar to fence searches except that they usually cover multiple rows of antenna beams in elevation. There is not a clear differentiation between volume and sector searches or between sector and horizon fence searches and these terms are often used synonymously. Sector search design is best performed using the radar range equation defined in Section 2.1 (i.e., treating sector search as a volume search) rather than as described in this section for the horizon fence search.

4.2.3 Cued Search

Cued search is a method primarily used to accomplish a handover from a sensor to a radar system. This type of search relies on a target state vector that includes at a minimum target position, rate, and a validity time. The position and rate are usually expressed in an inertial reference frame using Cartesian coordinates (e.g., $x, y, z, \dot{x}, \dot{y}, \dot{z}$). The size of the search volume can be based upon the dimensions of the predicted error covariance matrix associated with the target state vector when it is supplied with the state vector. The search volume must be transformed into the radar measurement coordinates (e.g., spherical: range, azimuth, and elevation or, equivalently, sine space coordinates). When the covariance matrix is not available, fixed range and angular extents can be used.

Since this type of search is based on tracking data, when the handover is radar-to-radar, track-bandwidth waveforms are typically used. In general, since these searches are small relative to other search types, they require fewer antenna beams to cover target position and rate uncertainties. However, since wider bandwidth waveforms are employed resulting in higher range resolution, the number of range cells to be searched can still be large. This can lead to relatively high signal processing load for matched filtering and subsequent detection processing.

The radar range equation for cued search is identical to that used for volume search. However, the cued search scan or frame time is driven by a different requirement than other searches. Horizon fence and volume searches are typically autonomous capabilities, that is, in general they have little or no prior information on likely target location. Therefore these searches must achieve their required cumulative probability of detection ($P_{D\ cum}$) before targets fly though coverage. Hence, scan time for volume and fence searches is driven to accomplish multiple looks in the time that a specified target will be in coverage.

Since cued searches are referenced (i.e., centered) to a predicted target state vector, presumably with an error covariance matrix or the equivalent information, the target is much less likely to fly out of coverage. Therefore, longer scan times can potentially be used for the much smaller search volumes. This can dramatically reduce the power-aperture product needed to implement cued searches.

Consider a cued search that must cover a 5-degree azimuth by 5-degree elevation solid angle (i.e., an approximation to the 3σ error ellipse for the handover). Assume that due to prior knowledge of target parameters (from the state vector and error covariance matrix) that a frame time of 1 second is adequate to ensure target capture in the search coverage. Table 4.3 illustrates the volume search range equation for this problem.

The result from Table 4.3 indicates that a power-aperture product of 34.95 dBW-m^2 is required to implement the cued search. Again, for a 10 m^2 antenna aperture, the required average power is 24.96 dBW or, equivalently, 313.17 W. For a 25% duty factor, this corresponds to a peak transmitter power of 1.25 kW. Comparing this result to the power-aperture products necessary for the volume and horizon fence searches of Sections 2.1 and 2.2, respectively, illustrates the lower radar resources potentially needed for cued searches.

For cued searches, the verification waveforms, if used, may be of identical bandwidth to the search waveform. Therefore, the transition from cued search to track typically may not require a separate acquisition or track initiation waveform if monopulse measurements are available with the verification waveforms. This is not usually the case for volume and horizon fence searches as is described in Section 3.

Table 4.3   Cued Search Design Example

Radar parameter	+	−
4π	10.99	
kT_s		200.81
R^4 (500 km)	227.95	
Ψ (5 degrees by 5 degrees)		21.18
L_t (3dB)	3.00	
L_r (3dB)	3.00	
SNR (12 dB per pulse)	12.00	
σ (1 meter^2)	0.00	
T_{SC} (1 second)		0.00
Totals	256.94	221.99

4.2.4 Multiple-Beam Search

For large search volumes and very small antenna beamwidths, the resulting situation is a large number of antenna beam positions to be serviced in a limited frame or scan time. This condition, which can occur for large antenna apertures at high operating frequencies, can stress the radar's timeline usage or occupancy. Based on the fraction of radar resources allocated to search (i.e., radar duty factor), a maximum available search pulse repetition frequency (PRF) can be calculated.

Consider an example where the radar can service 25 beams per second, and 50 percent of the radar is allocated to search. This situation results in 12.5 beams/second available for search. If the required search volume requires a beam-rate in excess of 12.5 per second, the radar is considered to be "occupancy-limited."

When radars are occupancy-limited, multiple simultaneous receive beams can be used to reduce the occupancy by the ratio of approximately $1/N_B$, where N_B is the number of simultaneous receive beams. Therefore, in the example cited above, if the search requires a beam-rate of 37.5 per second and 3 simultaneous receive beams are used, then the available 12.5 beams per second will satisfy the search requirements. The trade-off involved when using this approach is the need to provide N_B receiver channels, which can represent a cost impact to the radar design.

For monopulse tracking radars, the need for $N_B \leq 3$ is not usually an issue due to the availability of the two monopulse difference channels that are normally unused during most searches which use only the sum channel (the exception possibly being cued search). However, when $N_B > 3$, this requires the cost of $(N_B - 3)$ additional receiver channels. In the case of digital beam forming (DBF), this does not pose an additional cost since DBF by definition provides for multiple-beam operation (i.e., multiple receivers are already required to implement DBF). Normally, the simultaneous beams are formed at different operating frequencies to mitigate cross-talk between spatially-adjacent beams.

As for other search variants, some form of verification waveform is used prior to track initiation processing. Like the volume and horizon fence searches, all-range digital pulse compression is usually employed due to the large range extents.

4.3 ACQUISITION WAVEFORMS AND PROCESSING

Acquisition, also referred to as track initiation (TI), is the transition phase between the search and tracking functions. Since track usually uses wider bandwidth than most autonomous searches (e.g., volume or horizon fence searches), the purpose of this function is to obtain accurate initial estimates of target position and rate to commence the tracking process. More specifically, due to the recursive nature of most tracking algorithms (i.e., tracking filters), an initial state vector and error covariance matrix are necessary to begin the tracking process. Continuation of the tracking process after track initiation is referred to as track maintenance (TM). This is described in Chapter 5.

To maximize compatibility with track maintenance, the acquisition or TI waveforms are generally identical to those used for tracking. In addition, both range and monopulse processing of TI returns are performed to enable initial estimates of all state vector components. Since the state vector and error covariance is usually expressed in a Cartesian coordinate frame (i.e., $x, y, z, \dot{x}, \dot{y}, \dot{z}$), this generally requires transformation from radar measurement coordinates (e.g., a spherical coordinate system) into a Cartesian coordinate frame. Once the initial target state vector and error covariance matrix has been computed, these quantities are updated via the TM process.

4.4 REFERENCES

[1] D. K. Barton, *Radar System Analysis and Modeling*, Artech House, 2004

[2] E. Brookner, *Aspects of Modern Radars*, Artech House, 1988

[3] E. Brookner, *Practical Phased Array Antenna Systems*, Artech House, 1991

[4] J. DiFranco & W. Rubin, *Radar Detection*, SciTech, 2004

[5] N. Levanon, *Radar Principles*, Wiley-Interscience, 1988

[6] R. Nitzberg, *Radar Signal Processing and Adaptive Systems*, 2nd Edition, Artech House, 1999

[7] A. Papoulis, *Probability, Random Variables, and Stochastic Processes*, McGraw-Hill, 1965

[8] A. Papoulis, *Signal Analysis*, McGraw-Hill, 1977

[9] M. Skolnik, *Introduction to Radar Systems*, 3rd Edition, McGraw-Hill, 2002

[10] M. Skolnik, *Radar Handbook*, 2nd Edition, McGraw-Hill, 1990

[11] H. Van Trees, *Detection, Estimation and Modulation Theory, Part 1*, Wiley-Interscience, 2001

[12] D. K. Barton, *Modern Radar System Analysis*, Artech House, 1988

[13] J. V. Candy, *Signal Processing—The Modern Approach*, McGraw-Hill, 1988

[14] G. Minkler & J. Minkler, *CFAR*, Magellan, 1990

4.5 PROBLEMS

1. Consider an X-band phased-array radar that must perform three types of searches:

- Volume search:

  - Range: 1,000 km
  - Azimuth extent: ± 20 degrees
  - Elevation extent: 0 to 15 degrees
  - Target RCS: 0 dBsm
  - Required SNR: 12 dB
  - Scan time: 7.5 seconds

- Horizon fence:

  - Range: 2,000 km
  - Azimuth extent: ± 45 degrees

- Target RCS: +10 dBsm
- Required SNR/look: 8 dB
- Number of looks: 6
- Target vertical rate: 2.7 km/second

• Cued search:

- Range: 750 km
- Azimuth extent: ± 5 degrees
- Elevation extent: ± 5 degrees
- Target RCS: −10 dBsm
- Scan time: 2 seconds.

Assume that the system noise temperature is 500 kelvins and that the total transmit and receive losses are 4.5 dB. Calculate the required average power-aperture product for all three search modes. Which search mode sizes the radar?

2. Consider an S-band radar (λ = 0.09 meter) sized to perform a volume search of 65 degrees in azimuth by 65 degrees in elevation in 2 seconds. If the unweighted antenna aperture is circular with an effective area 1 meter^2, the radar can provide a maximum PRF of 40 beams per second, and 60 percent of the radar is allocated to performing the search function, determine if the radar is occupancy limited. If so, approximately how many receiver channels are required to implement the search using multiple-simultaneous receive beams?

5

Estimation, Tracking, and Data Association

5.1 INTRODUCTION

This chapter covers the concepts of parameter estimation, target tracking, and data association algorithms used to enable multiple-target tracking in real-world environments. References [1] through [4] are excellent sources of additional information on tracking. The topics covered in this chapter include:

- Parameter estimation for radar
- The radar tracking function:
  - Waveforms and signal processing
- Types of tracking filters:
  - Alpha-beta and alpha-beta-gamma
  - Kalman
  - Extended-Kalman
  - Interacting multiple-model
- Data association algorithms:
  - Nearest-neighbor
  - Probabilistic data association (PDA)

- Joint PDA (JPDA)
- Nearest neighbor-JPDA
- Multi-hypothesis tracking (MHT)
- Other assignment algorithms

• Tracking air targets:

- Aircraft, unmanned aerial vehicles (UAVs)
- Cruise missiles

• Tracking ballistic missile targets:

- TBMs, IRBMs, ICBMs

• Tracking surface targets:

- Ships
- Vehicles

5.2 PARAMETER ESTIMATION FOR RADAR

Parameter estimation is an important function performed by radars. It differs from radar measurements or observations, inasmuch as that in this chapter estimation refers to extracting parameters not directly measured by radar. The most common applications of estimation in radar are target tracking and target feature estimation to enable target classification and discrimination. This chapter focuses on estimation in the target tracking context, while Chapter 6 treats target feature estimation for classification, discrimination, and identification.

References [4], [5], and [6] are three sources of theory related to parameter estimation. In general, parameter estimation solves the problem of extracting desired quantities from radar measurements, that is:

$$\hat{x} = E\left\{ x \middle| z \right\},\tag{5.1}$$

where x is the parameter being estimated and z are observations from which estimates of x can be extracted. It is assumed that x is observable in the control theory sense, that is, that quantities necessary to estimate x are included in a linear fashion in measurement z.

There are three important and desirable characteristics of parameter estimates and estimators. First, the estimate should be unbiased, that is, the estimation error should have a mean value of zero. Second, the variance of the estimation error should be reasonably small or at least understood. Equations (5.2) and (5.3) are mathematical representations of these concepts:

$$E\{ x - \hat{x} \} = E\{ \tilde{x} \} = 0, \tag{5.2}$$

and

$$E\{ (x - \hat{x})^2 \} = E\{ \tilde{x}^2 \} = \sigma_{\tilde{x}}^2. \tag{5.3}$$

A third desirable characteristic of an estimator is that the variance of the estimate approaches its theoretical lower bound asymptotically as more measurements are available. Estimators that achieve this property are referred to as efficient estimators.

There exists a theoretical lower bound on the value of estimation error variance referred to as the Cramer-Rao bound. When N statistically independent measurements of x are averaged with a given signal-to-noise ratio (SNR) per observation, the Cramer-Rao bound on the error variance of the estimate is approximately given by:

$$\sigma_{CRB}^2 = \frac{(\Delta x)^2}{N \; SNR}, \tag{5.4}$$

where Δx is the measurement resolution of x and N is the number of statistically independent measurements of x. As will be seen, the lower bound in equation (5.4) is important for predicting estimation performance, and specifically for predicting tracking accuracy.

5.3 THE RADAR TRACKING FUNCTION

The radar tracking function is illustrated using the block diagram in Figure 5.1. As can be seen, the antenna is steered in the direction of the predicted target po-

```
┌─────────────────┐     ┌─────────────────┐     ┌─────────────────┐
│  Steer Antenna  │     │ Transmit Track  │     │   Associate     │
│  at Predicted   │ ──▷ │  Waveform &     │ ──▷ │     Radar       │
│ Target Position │     │ Receive Target  │     │  Measurements   │
│                 │     │     Echo        │     │  with Existing  │
│                 │     │                 │     │  Target Tracks  │
└─────────────────┘     └─────────────────┘     └─────────────────┘
        △                                                │
        │                                                ▽
        │               ┌─────────────────┐     ┌─────────────────┐
        │               │ Schedule Track  │     │   Smooth &      │
        └────────────── │  Maintenance    │ ◁── │ Predict Track   │
                        │   Waveform      │     │  State Vectors  │
                        └─────────────────┘     └─────────────────┘
```

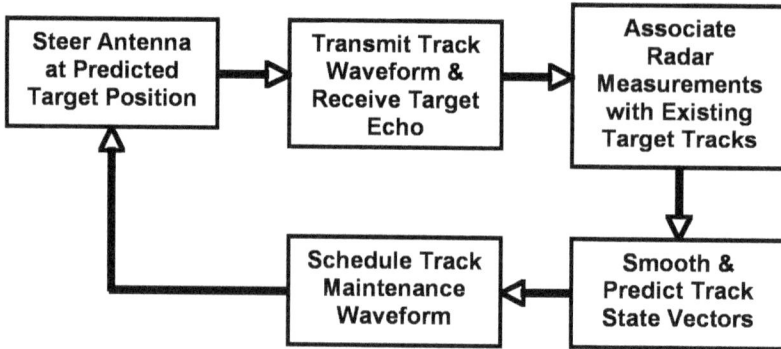

Figure 5.1 Block Diagram of the Radar Tracking Function

sition, track waveforms are transmitted to the target, target echoes are received, these measurements are associated with current targets, the target state vectors are updated using the new measurement, the state vector is predicted to the time of the next track update, and this information is used to steer the antenna starting the process again. This functional sequence is referred to as a tracking loop, and executing it is referred to as closing the track loop.

The two major sub-functions that comprise the radar tracking function are the data association algorithm and the tracking filter. The remaining functions in Figure 5.1 are performed by the radar scheduler (schedule the track waveform and steer the antenna), the radar transmitter (transmit the track waveform), and the receiver (receive the target echo).

The waveforms and signal processing associated with the tracking function are as follows. First, narrowband waveforms are used for tracking, that is, radio frequency bandwidths in the range of 5 to 50 MHz. The signal processing performed is the standard pulse matched filtering, followed by range and amplitude interpolation and peak detection. Monopulse processing is used to estimate target sine space angles u and v. For cases of low-elevation tracking, multipulse waveforms and processing are employed (e.g., MTI or pulse-Doppler) to mitigate clutter backscatter effects.

The subsequent sections in this chapter are devoted to describing the tracking filters, data association algorithms, and the specific tracking of air, ballistic missile, and surface targets. However, first the subject of the coordinate systems and transformations between them will be discussed briefly in the next section.

5.3.1 Coordinate Systems

Coordinate systems are an important aspect of the radar tracking function. Since phased-array radars are the focus of this book, the measurement coordinate system is a special form of spherical coordinates: range and the two direction-cosines u and v, referenced to the antenna array boresight, referred to as (R, u, v) coordinates.

However, most tracking systems use an inertial frame of reference to estimate target position, typically expressed in Cartesian coordinates: (x, y, z). This reference system is selected since most target classes move most naturally in Cartesian coordinates, not in spherical coordinates. Therefore, at a minimum, transformations between (R, u, v) and (x, y, z) are necessary to relate measurements to the target states used by the tracking filters. These effectively convert the radar measurements to Cartesian array face coordinates, and then translate and rotate them to Cartesian inertial coordinates. For fixed-antenna arrays on stationary platforms, this is all that is required.

The situation is more complicated for moving antenna arrays, whether mounted on pedestals or mounts, or on a moving platform such as a ship, aircraft, or missile. In these cases additional rotations and translations are needed to account for the pedestal and/or platform motion relative to the Cartesian inertial frame in which target tracks are updated and predicted.

Therefore, a sequence of coordinate transformations and translations are a fundamental component of the radar tracking process. In most cases, Cartesian inertial coordinates will be used in discussing the tracking algorithms throughout the remainder of this chapter, with the exception of data association algorithms, which must relate measurements to state vectors. However, it is important to not lose sight of these coordinate systems when designing and analyzing radar tracking systems.

5.4 TYPES OF TRACKING FILTERS

5.4.1 Fixed-Gain Filters

There are many types of tracking filters in use for radar tracking applications. However, there are two fundamental classes of tracking filters: (i) fixed-gain filters and (ii) computed-gain filters. The first class is the simplest and its implementations use the least data processing throughput per tracked target. The most common of these are the $\alpha - \beta$ and $\alpha - \beta - \gamma$ or alpha-beta and alpha-beta-gamma filters. The mathematical forms are:

$$\begin{bmatrix} x \\ \dot{x} \end{bmatrix}_{k+1} = \left(z_{k+1} - \begin{bmatrix} x \\ \dot{x} \end{bmatrix}_{k} \right)^{T} \begin{bmatrix} \alpha_{k+1} & 0 \\ 0 & \beta_{k+1} \end{bmatrix} + \begin{bmatrix} x \\ \dot{x} \end{bmatrix}, \quad (5.5)$$

and

$$\begin{bmatrix} x \\ \dot{x} \\ \ddot{x} \end{bmatrix}_{k+1} = \left(z_{k+1} - \begin{bmatrix} x \\ \dot{x} \\ \ddot{x} \end{bmatrix}_{k} \right)^{T} \begin{bmatrix} \alpha_{k+1} & 0 & 0 \\ 0 & \beta_{k+1} & 0 \\ 0 & 0 & \gamma_{k+1} \end{bmatrix} + \begin{bmatrix} x \\ \dot{x} \\ \ddot{x} \end{bmatrix}, \quad (5.6)$$

where z is the measurement vector at the $k + 1$ sample time, and α, β, and γ are the fixed or precomputed weights for sample time $k + 1$. As can be seen, since only a small number of adds, subtracts, and multiplies are required for each track update, the computational requirements for these filters are small.

5.4.2   Computed-Gain Filters

5.4.2.1 Kalman Filters. There are many forms of computed-gain tracking filters. However, the most common and widely used is the Kalman filter (KF) type. These filters are in the statistical filtering class, that is, they embody dynamical models of the specific target motion and use these to propagate the expected-value of the state estimates and the covariance matrix of the estimation errors. The gain computations are where the Kalman filters incur the majority of necessary computations. The state vector update equations are very similar to those in equations (5.5) and (5.6).

The Kalman filter, references [1] through [4], is the optimal filter if the target dynamics and the measurement-to-state relationships are linear. If one or both of these relationships are nonlinear, then the Kalman filter is the optimal linear filter. Theoretically, an optimal nonlinear filter does exist; however, there is no systematic method to determine its form. Hence, this is the reason that some form of the Kalman filter is used in many if not most target tracking applications. In most cases, due to the non-linearities mentioned above, a special form of the Kalman filter is used in these cases. It is referred to as the Extended-Kalman filter (EKF). The EKF is a Kalman filter matched to linearized version of the equations of state dynamics and observations.

The fundamental concept behind the Kalman filter is the minimization of the estimation error in the mean-squared sense, and so the Kalman filter is often referred to as a minimum mean-squared estimate (MMSE). Figure 5.2 is a flowchart of the sequence of computations required to implement one track update and prediction of the target state and error covariance matrix.

It should be noted that the Kalman filter is a recursive estimator, that is, it updates its estimates upon receipt of each new measurement. To illustrate the idea of recursive estimation, consider the problem of recursive estimating the arithmetic mean of a sequence of values. Figure 5.3 shows the estimation sequence.

In a similar fashion the Kalman filter's recursive formulation can be derived. This is shown in Figure 5.4. This class of filter is often referred to as a predictor-corrector form, as indicated in Figure 5.4d.

Practical tracking issues that the Kalman filter does not handle include:

• Nonlinear motion models (use linearization, i.e., the extended KF [EKF])

• Nonlinear measurement equations (use EKF and a *de-biased consistent* transformation [polar-to-Cartesian])

• Unknown inputs to dynamic (system) equation and/or mode changes (different motion models, e.g., uniform vs. acceleration or turn)

• Correlated noises (auto and cross-correlated)

• Unknown sensor resolution and multipath propagation

• Unknown number of targets

• Unknown origin of measurements: data association uncertainty.

However, due to its optimality, simplicity, and systematic handling of correlated target dynamics and correlated measurement errors, and since these minor drawbacks can be mitigated by proven techniques, such as using the Extended-Kalman filter for nonlinear problems, addition of process noise (the filter's uncertainty of the true underlying dynamical models), increasing update rates to

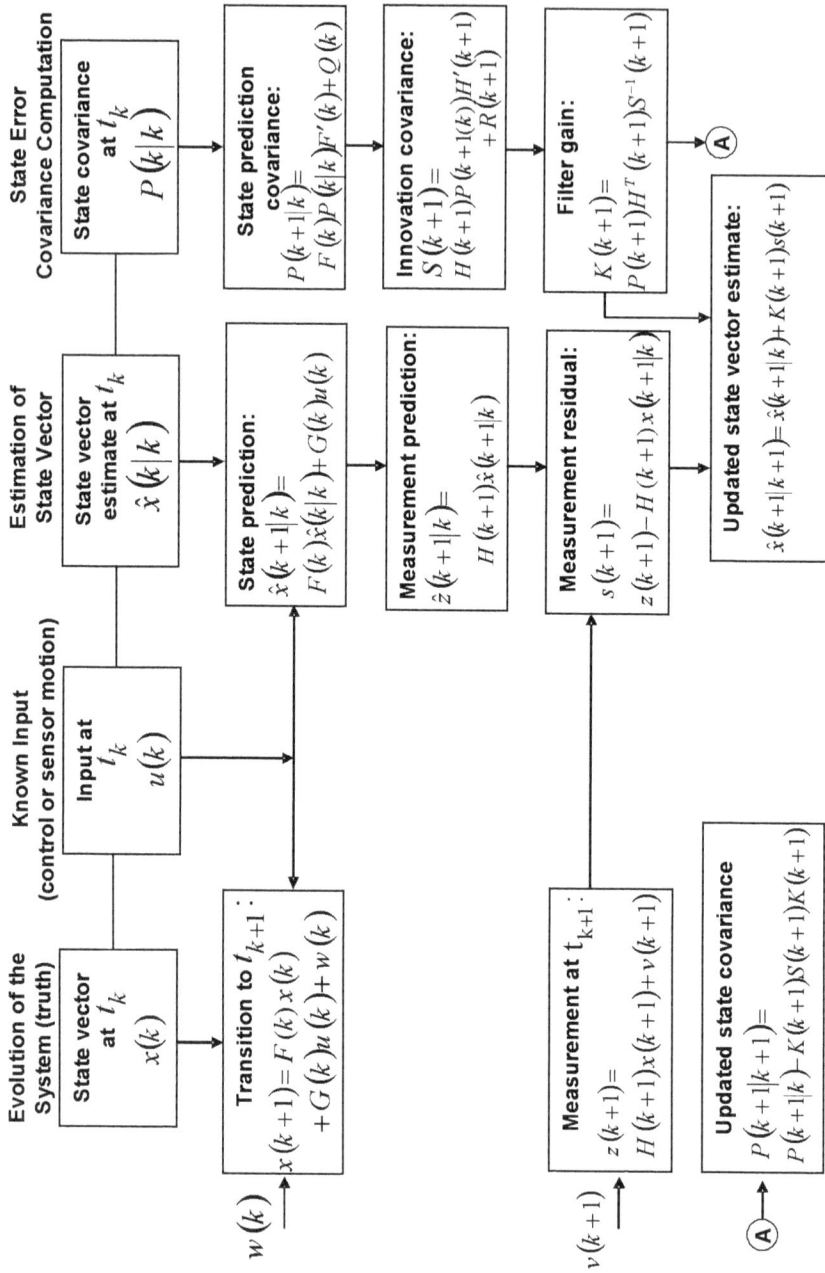

State Error Covariance Computation

State covariance at t_k
$$P(k|k)$$

State prediction covariance:
$$P(k+1|k)= F(k)P(k|k)F'(k)+Q(k)$$

Innovation covariance:
$$S(k+1)= H(k+1)P(k+1|k)H'(k+1) +R(k+1)$$

Filter gain:
$$K(k+1)= P(k+1)H^T(k+1)S^{-1}(k+1)$$

(A)

Estimation of State Vector

State vector estimate at t_k
$$\hat{x}(k|k)$$

State prediction:
$$\hat{x}(k+1|k)= F(k)\hat{x}(k|k)+G(k)u(k)$$

Measurement prediction:
$$\hat{z}(k+1|k)= H(k+1)\hat{x}(k+1|k)$$

Measurement residual:
$$s(k+1)= z(k+1)-H(k+1)x(k+1|k)$$

Updated state vector estimate:
$$\hat{x}(k+1|k+1)=\hat{x}(k+1|k)+K(k+1)s(k+1)$$

Known Input (control or sensor motion)

Input at t_k
$$u(k)$$

Evolution of the System (truth)

State vector at t_k
$$x(k)$$

$w(k) \longrightarrow$

Transition to t_{k+1}:
$$x(k+1)= F(k)x(k) +G(k)u(k)+w(k)$$

$v(k+1) \longrightarrow$

Measurement at t_{k+1}:
$$z(k+1)= H(k+1)x(k+1)+v(k+1)$$

(A)

Updated state covariance
$$P(k+1|k+1)= P(k+1|k)-K(k+1)S(k+1)K(k+1)$$

Figure 5.2 Kalman Filter Processing Sequence

Consider the simple problem of calculating an average:

$$Average(N) = \frac{1}{N}\sum_{k=1}^{N} a(k) \qquad \textbf{Batch estimate}$$

This can be expressed as:

$$
\begin{aligned}
Average(N) &= \frac{1}{N}\left[a(1) + a(2) + a(3) + \dots + a(N) \right] \\
&= \frac{1}{N}\left[\sum_{k=1}^{N-1} a(i) + a(N) \right] \\
&= \frac{1}{N}\left[\frac{N-1}{N-1} \sum_{k=1}^{N-1} a(i) + a(N) \right] \\
&= \frac{N-1}{N}\; Average(N-1) + \frac{1}{N}\; a(N) \qquad \textbf{Recursive estimate}
\end{aligned}
$$

Figure 5.3   Recursive Estimate of Arithmetic Mean

Define:

$$P_{k/k} = E\left\{\widetilde{x}_{k/k}\,\widetilde{x}_{k/k}^T\right\}$$

If the predicted state estimate is defined as:

$$\hat{x}_{k+1/k} = F_{k+1}\,x_k + w_{k+1}$$

Then the predicted state error covariance matrix is:

$$\begin{aligned}
P_{k+1/k} &= E\left\{\widetilde{x}_{k+1/k}\,\widetilde{x}_{k+1/k}^T\right\} \\
&= F_{k+1}\,E\left\{\widetilde{x}_{k/k}\,\widetilde{x}_{k/k}^T\right\}F_{k+1}^T + E\left\{w_{k+1}\,w_{k+1}^T\right\} \\
&= F_{k+1}\,P_{k/k}\,F_{k+1}^T + Q_{k+1}
\end{aligned}$$

If the next measurement is defined as:

$$z_{k+1} = H_{k+1}\,x_{k+1} + v_{k+1}$$

Then the predicted measurement error covariance matrix is:

$$E\left\{\widetilde{z}_{k+1}\,\widetilde{z}_{k+1}^T\right\} = E\left\{v_{k+1}\,v_{k+1}^T\right\} = R_{k+1}$$

Process noise covariance:

$$Q_{k+1} = E\left\{w_{k+1}\,w_{k+1}^T\right\}$$

Figure 5.4a Derivation of Recursive Kalman Filter Equations

$$\hat{x}_{k+1/k} = F_k\,\hat{x}_{k/k} + w_{k+1}$$

Define the predictor-corrector filter as:

$$\hat{x}_{k+1/k+1} = \hat{x}_{k+1/k} + K_{k+1}\left[z_{k+1} - H_{k+1}\,\hat{x}_{k+1/k}\right]$$
$$= \hat{x}_{k+1/k} + K_{k+1}\,s_{k+1}$$

Define the innovation as:

$$s_{k+1} = \left[z_{k+1} - H_{k+1}\,\hat{x}_{k+1/k}\right]$$

Define state estimation error covariance as:

$$E\left\{\tilde{x}_{k+1/k+1}\,\tilde{x}_{k+1/k+1}^T\right\} = E\left\{\left[\tilde{x}_{k+1/k} + K_{k+1}\,\tilde{s}_{k+1}\right]\left[\tilde{x}_{k+1/k} + K_{k+1}\,\tilde{s}_{k+1}\right]^T\right\}$$

$$= E\left\{\tilde{x}_{k+1/k}\,\tilde{x}_{k+1/k}^T\right\} + E\left\{K_{k+1}\,\tilde{s}_{k+1}\,\tilde{x}_{k+1/k}^T\right\} + E\left\{\tilde{x}_{k+1/k}\,\tilde{s}_{k+1}^T\,K_{k+1}^T\right\}$$
$$+ E\left\{K_{k+1}\,\tilde{s}_{k+1}\,\tilde{s}_{k+1}^T\,K_{k+1}^T\right\}$$

$$= P_{k+1/k} - K_{k+1}\,H_{k+1}\,E\left\{\tilde{x}_{k+1/k}\,\tilde{x}_{k+1/k}^T\right\} - E\left\{\tilde{x}_{k+1/k}\,\tilde{x}_{k+1/k}^T\right\}H_{k+1}^T\,K_{k+1}^T$$
$$+ K_{k+1}\,E\left\{\tilde{s}_{k+1}\,\tilde{s}_{k+1}^T\right\}K_{k+1}^T$$

$$= P_{k+1/k} - K_{k+1}\,H_{k+1}\,P_{k+1/k} - P_{k+1/k}\,H_{k+1}^T\,K_{k+1}^T + K_{k+1}\,S_{k+1}\,K_{k+1}^T$$

Figure 5.4b Derivation of Recursive Kalman Filter Equations (continued)

Define state estimation error covariance as:

$$E\left\{\widetilde{x}_{k+1/k+1}\;\widetilde{x}_{k+1/k+1}^{T}\right\} = P_{k+1/k} \;-\; K_{k+1}\,H_{k+1}\,P_{k+1/k} \;-\; P_{k+1/k}\,H_{k+1}^{T}\,K_{k+1}^{T} \;+\; K_{k+1}\,S_{k+1}\,K_{k+1}^{T}$$

Find the filter gain that minimizes the state error covariance:

$$\frac{\partial E\left\{\widetilde{x}_{k+1/k+1}\;\widetilde{x}_{k+1/k+1}^{T}\right\}}{\partial K_{k+1}^{T}} = 0 = 0 \;-\; 2P_{k+1/k}\,H_{k+1}^{T} \;+\; 2K_{k+1}\,S_{k+1}$$

The optimum filter gain is then:

$$K_{k+1} = P_{k+1/k}\,H_{k+1}^{T}\,S_{k+1}^{-1}$$

$$\begin{aligned}
S_{k+1} &= E\left\{\widetilde{s}_{k+1}\;\widetilde{s}_{k+1}^{T}\right\} \text{ Innovation covariance matrix}\\
&= E\left\{\widetilde{z}_{k+1}\;\widetilde{z}_{k+1}^{T}\right\} + H_{k+1}\,E\left\{\widetilde{x}_{k+1/k}\;\widetilde{x}_{k+1/k}^{T}\right\}H_{k+1}^{T}\\
&= E\left\{v_{k+1}\;v_{k+1}^{T}\right\} + H_{k+1}\,E\left\{\widetilde{x}_{k+1/k}\;\widetilde{x}_{k+1/k}^{T}\right\}H_{k+1}^{T}\\
&= R_{k+1} + H_{k+1}\,P_{k+1/k}\,H_{k+1}^{T}
\end{aligned}$$

Substituting for the innovation covariance matrix yields the gain:

$$K_{k+1} = P_{k+1/k}\,H_{k+1}^{T}\left[R_{k+1} + H_{k+1}\,P_{k+1/k}\,H_{k+1}^{T}\right]^{-1}$$

Figure 5.4c Derivation of Recursive Kalman Filter Equations (continued)

The updated state estimation error covariance matrix then becomes:

$$P_{k+1/k+1} = P_{k+1/k} - K_{k+1} S_{k+1} K_{k+1}^T$$

$$= P_{k+1/k} - K_{k+1} \left[R_{k+1} + H_{k+1} P_{k+1/k} H_{k+1}^T \right] K_{k+1}^T$$

Equivalently:

$$P_{k+1/k+1} = P_{k+1/k} - P_{k+1/k} H_{k+1}^T S_{k+1}^{-1} S_{k+1} S_{k+1}^{-1} H_{k+1} P_{k+1/k}$$

$$= P_{k+1/k} - P_{k+1/k} H_{k+1}^T S_{k+1}^{-1} H_{k+1} P_{k+1/k}$$

$$= P_{k+1/k} - K_{k+1} H_{k+1} P_{k+1/k}$$

$$= \left[I - K_{k+1} H_{k+1} \right] P_{k+1/k}$$

$$P_{k+1/k+1} = \underbrace{\left[I - K_{k+1} H_{k+1} \right]}_{\substack{\text{Corrector} \\ \text{Improvement}}} \underbrace{\left[F_{k+1} P_{k/k} F_{k+1}^T + Q_{k+1} \right]}_{\substack{\text{Predictor} \\ \text{Growth}}}$$

Figure 5.4d Derivation of Recursive Kalman Filter Equations (continued)

ameliorate residual nonlinear effects, and so on, it is the ubiquitous "workhorse" of tracking filters.

5.4.2.2 Interacting Multiple-Model Filters. Interacting multiple-model (IMM) filters consist of a bank of parallel Kalman filters that implement different target dynamical models. These are blended or fused based on a Bayesian approach as shown in Figure 5.5. The IMM filters provide the capability for a tracking filter to use different target models during a target's trajectory to minimize tracking errors. Unlike tracking approaches like MHT, which propagate all possible target hypotheses at each track update, the IMM only propagates a single step at each update. This yields a computational load of approximately M times that of a sin-

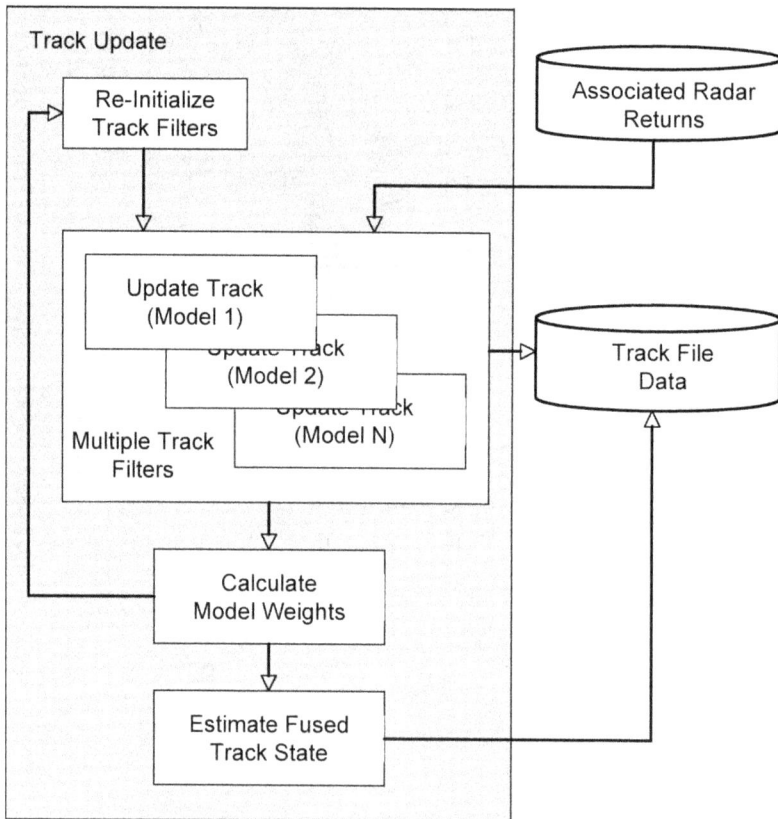

Figure 5.5 IMM Filter Block Diagram

gle Kalman filter, where M is the number of target models employed. The detailed filtering logic is described in the following paragraphs.

The steps of the IMM approach are given below.

1. Model Conditioned Reinitialization

    Each EKF is reinitialized with a composite state and covariance whose constituent components are weighted by the conditional probability that the target is transitioning to the particular model.

    The conditional probability that the target is transitioning to (or remains as) model i is:

    $$\mu_{j|i}^{(k-1)} = \frac{\pi_{ji}\mu_j^{(k-1)}}{\mu_i^{(k,k-1)}} \text{ , where}$$

    $$\mu_i^{(k,k-1)} = \sum_{j=1}^{M} \pi_{ji}\mu_j^{(k-1)} \text{ is the total probability of being in model } i,$$

    π_{ij} is the a priori Markov transition probability that the target is transitioning from model i to model j, and M is the number of models.

    The states and covariance matrices are mixed using the conditional probabilities for each respective model as weights.

    $$\bar{x}_i^{(k-1,k-1)} = \sum_{j=1}^{M} \mu_{j|i}^{(k-1)}\bar{x}_j^{(k-1,k-1)} \text{ , is the state used to reinitialize the } i^{th} \text{ model.}$$

    $$P_i^{(k-1,k-1)} = \sum_{j=1}^{M} \mu_{j|i}^{(k-1)}\left[P_i^{(k-1,k-1)} + \Delta P_{ji}^{(k-1,k-1)} \right] \text{, is the covariance matrix used}$$

to reinitialize the ith model, where:

$$\Delta P_{ij} = (\bar{x}_i - \bar{x}_j)(\bar{x}_i - \bar{x}_j)^T \text{ is added to account for the differences between}$$

models i and j.

2. Model-Conditioned Filtering and Prediction

   Update the state and covariance of all EKFs using the new measurement $\vec{z}_k$ for the $(k+1)^{\text{th}}$ innovation.

3. Model Probability Update

   The composite state and covariance matrix are blended based on the posterior probability for each model, given below:

$$\mu_i^{(k)} = \frac{L_i^k \mu_i^{(k,k-1)}}{\sum_{j=1}^{M} L_j^k \mu_j^{(k,k-1)}} \quad \text{where,}$$

$$\mu_i^{(k,k-1)} = \sum_{j=1}^{M} \pi_{ji} \mu_j^{(k-1)} \quad \text{is the total probability of model } i \text{ calculated in step 1.}$$

$$L_i = \frac{1}{\sqrt{(2\pi)^m |S_i|}} \exp(-\xi_i^2 / 2) \quad \text{is the likelihood of the target being in model}$$

i, and

$$\xi_i^2 = \vec{y}_i^T S_i^{-1} \vec{y}_i \quad \text{is the chi-square statistical distance of the measured state}$$

from the predicted state.

$$\vec{y} = \left[\vec{z}_k - H_k(\vec{x}_{k,k-1}) \right],$$

$$S_i = \left[H_k P_{k,k-1} H_k^T + R \right],$$

where m is the dimension of the state.

4. Estimate Fusion

   This step produces the output of the tracking system used to determine the associated measurement.

$$\bar{\mathbf{x}}^{(k,k)} = \sum_{i=1}^{M} \mu_i^{(k)} \bar{\mathbf{x}}_i^{(k,k)} \text{ , is the state used to reinitialize the } i^{\text{th}} \text{ model.}$$

$$\mathbf{P}^{(k,k)} = \sum_{i=1}^{M} \mu_i^{(k)} \left[\mathbf{P}_i^{(k,k)} + \Delta\mathbf{P}_i^{(k,k)} \right] \text{ is the covariance used to reinitialize the } i^{\text{th}}$$

model, where the term:

$$\Delta\mathbf{P}_i = (\bar{\mathbf{x}} - \bar{\mathbf{x}}_i)(\bar{\mathbf{x}} - \bar{\mathbf{x}}_i)^{\mathrm{T}} \text{ corrects the covariance matrix for the difference be-}$$
tween the fused estimate and model i.

The output of the IMM tracking system is the composite state and covari-

ance $\left[\bar{\mathbf{x}}^{(k,k)}, \mathbf{P}^{(k,k)} \right]_{Composite}$, which are used in the association algorithm to de-

termine the measurement for the next innovation.

5.5   DATA ASSOCIATION ALGORITHMS

There are two fundamental classes of data association algorithms: (i) Non-Bayesian, and (ii) Bayesian approaches. These will be discussed in the following sections.

5.5.1   Nearest Neighbor

The "Achilles heel" of tracking is the correct association of new radar measurements to their originating target. Good data association, therefore, is the necessary condition for achieving good target tracking performance. The data association problem is especially challenging in environments with residual clutter echoes, densely spaced targets or target complexes, missed detections, and high false alarm rates. Therefore, the type of data association algorithm needed for a given tracking problem depends on these factors.

In benign situations with a few widely spaced targets, very high probability of detection, very low probability of false alarm, in a clear environment (i.e., no clutter, jamming, or other interference), almost any data association algorithm can be made to operate adequately. If these conditions are "guaranteed" to be the case, then a simple data association algorithm can and should be used. One such non-Bayesian algorithm is the nearest-neighbor (NN) technique. This algorithm updates each track with the new target detection closest to it (in the statistical "distance" sense). The NN algorithm can be expressed mathematically as:

$$\frac{\left(R_{track} - R_{measured}\right)^2}{\sigma_R^2} + \frac{\left(\theta_{track} - \theta_{measured}\right)^2}{\sigma_\theta^2} + \frac{\left(\phi_{track} - \phi_{measured}\right)^2}{\sigma_\phi^2} \leq D, \quad (5.7)$$

where the sum of normalized squared-errors in range, azimuth angle, and elevation angle, for example, is compared to a threshold and the track yielding the smallest of those sums for a given target measurement vector compared to the track state vector quantities would be updated using that measurement. This process is repeated for all measurements.

If the sum in equation (5.7) exceeds the threshold, D, then no measurement-to-track assignment is made, since exceeding D corresponds to a very small probability of correct target-to-track association. If the errors (differences in the numerators) in equation can be modeled as zero-mean Gaussian random variables, then the sum results in a chi-square distribution with three degrees-of-freedom. This allows D to be calculated to edit associations with any arbitrary probability of false association. When using a Kalman-type filter, the estimation error variances in the denominators can be obtained from the filter's computed error covariance matrix.

5.5.2   Probabilistic Data Association

The Probabilistic Data Association (PDA) algorithm uses a computed likelihood or estimated association probabilities to aid in solving the return-to-track problem rather than a simple proximity rule as for the NN algorithm. It is a Bayesian approach to the data association problem. Although it represents a higher computational load, it provides superior association performance in the presence of densely spaced targets (or clutter, etc.) to the NN method.

At each step in the association process, the PDA computes probabilities of correct measurement-to-track association for each measurement-track pair. The pair with the highest probability determines the assignment of that measurement. This process is repeated for all measurements. The detailed equations and processing logic are provided in detail in references [1] through [3] and will not be repeated here.

5.5.3   Joint Probabilistic Data Association

The Joint Probabilistic Data Association (JPDA) algorithm also uses a computed likelihood or estimated association probabilities to aid in solving the return-to-track problem rather than a simple proximity rule as for the NN algorithm. It is again a Bayesian approach to the data association problem. It represents a higher computational load compared to the simpler PDA but provides superior association performance in the presence of densely-spaced targets (or clutter, etc.) to the PDA method.

At each step in the association process, the JPDA computes probabilities of correct measurement-to-track association for each measurement-track pair considering all possible pairings in a joint-probabilistic sense. Similar to the PDA approach, the highest probability determines the assignment of measurements. The detailed equations and processing logic for JPDA are also provided in detail in references [1] through [3] and are not provided here.

5.5.4   Nearest-Neighbor JPDA

This technique developed by R. Fitzgerald at Raytheon is a combination of the NN and JPDA as its name suggests. It involves preprocessing the measurements using the NN to edit out unlikely measurement-track pairs. The JPDA algorithm is then applied to those pairs surviving this screening process. This can yield excellent performance at much lower computational cost than JPDA alone. Many phased-array radars use this data association technique. It provides association performance that can approach that of the more theoretically optimal multiple-hypothesis track (MHT) association technique in certain situations, while being much less computationally-intensive than MHT methods. MHT is briefly discussed next.

5.5.5 Multiple-Hypothesis Track

Multiple-hypothesis track is also a Bayesian approach to data association. Unlike the previously described methods that compare measurement-track pairs using statistical distance or likelihood measures, MHT creates and maintains a history of all possible data association hypotheses, creating a new and larger set of hypotheses at each track update time. As can be imagined, if no "hypothesis pruning" is performed, the number of hypotheses maintained grows exponentially as a function of time. Hence, use of a poorly designed MHT approach can have severe computational load consequences.

The source of MHT's optimality is also the source of its computationally intense nature. Since, in the unconstrained MHT approach, all possible measurement-to-track hypotheses are carried along, the correct hypothesis is available to the algorithm (the issue is selecting the correct one) and it can therefore be considered to be theoretically near-optimal. However, unless some pruning of unlikely hypotheses is performed, MHT is not a practical solution to the data association problem.

That being said, MHT, with suitable pruning logic, has been successfully used for many tracking applications. [3] is an excellent source to learn more about MHT and its practical applications.

5.5.6 Other Assignment Algorithms

Although the algorithms described in Sections 5.1 through 5.5 are commonly implemented techniques for data association, there exist many other approaches, and more continue to be developed. Table 5.1 is a partial listing of current techniques, including other data association algorithms other than those previously discussed compiled by F. Daum [3, 8, 9, 10].

As can be seen, the algorithms are rated based on a number of performance criteria, including:

• Ability to handle unresolved measurement data

• Performance in densely spaced target (e.g., clutter) environments

• Computational complexity (i.e., computer throughput usage).

Table 5.1 Partial List of Data Association Algorithms, after [3, 8, 9, 10]

Algorithm	Time horizon Considered (no. of Samples)	Number of Data Association Hypotheses	Unsolved Data Modeled in Algorithm	Relative Performance in Dense Multiple Target Environments		Computational Complexity	
				Unresolved Data	Resolved Data	Exact Solution	Approximate Solution
Nearest neighbor	1	1	No	Poor	Poor	Low	Low
Probabilistic data association (PDA)	1	1	No	Poor	Fair	Low	Low
Joint probabilistic data association (JPDA)	1	1	No	Fair	Fair	Poly	Medium
Nearest-neighbor JPDA	1	1	No	Fair	Good	Poly	Medium
Assignment	1	1	No	Fair	Good	Poly	Medium
Multidimensional assignment	N	1	No	Good	Excellent	Poly	High
Multiple hypothesis tracker (MHT)	N	Many	No	Good	Excellent	Exp	High
Koch MHT	N	Many	Yes!	Excellent!	Excellent	Exp	High

5.6 TRACKING AIR TARGETS

Tracking air targets poses several challenges. This is in part due to the variety of target types contained within the air target class, including:

• Manned aircraft

• Unmanned aerial vehicles (UAVs)

• Helicopters

• Cruise missiles.

Tracking manned aircraft is problematic compared with tracking ballistic missiles for the reason that the pilot in the aircraft can choose to maneuver at any time, and can select a number of distinct types of maneuvers. The ballistic missile after rocket burnout is effectively traveling in a fixed-parameter parabolic trajectory with only gravity acting upon it until it reenters the earth's atmosphere and is subject to drag deceleration.

UAVs may fly trajectories similar to a manned aircraft or not depending on the UAV capabilities and the control strategy. Similarly, there are several types of cruise missiles, some that fly at very low altitudes at subsonic speeds and can be terrain-following, and some that can be launched from high altitudes, travel at supersonic speeds, and can dive at steep angles when attacking.

For these reasons, a single target model cannot be used for tracking the broad class of air targets. The IMM filter described in Section 4.2.2 is one method to accommodate a broad range of possible target dynamics. Consider the following distinct behaviors for air targets:

• Constant-speed, straight and level flight

• Constant-acceleration, straight and level flight

• 3-G turn in plane

• 3-G climb or dive.

Possible state transitions (Figure 5.6) might be:

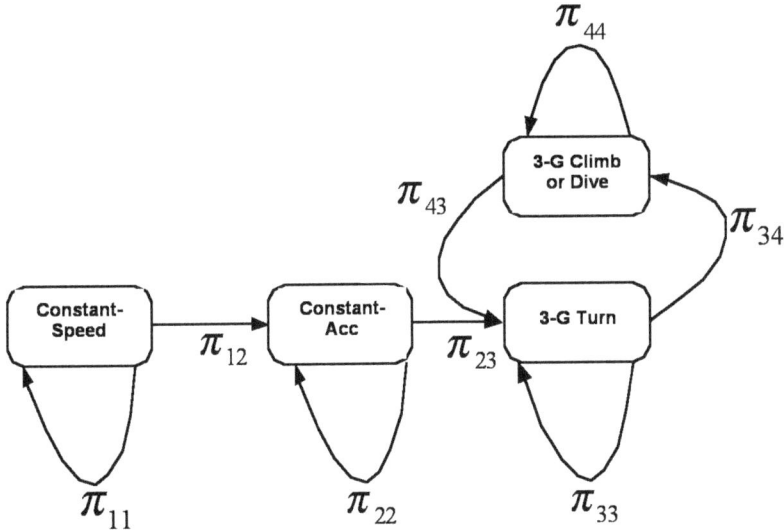

Figure 5.6 Possible Air Target State Transitions

$$\Pi = \begin{bmatrix} \pi_{11} & \pi_{12} & 0 & 0 \\ 0 & \pi_{22} & \pi_{23} & 0 \\ 0 & 0 & \pi_{33} & \pi_{34} \\ 0 & 0 & \pi_{43} & \pi_{44} \end{bmatrix}, \text{where}$$

π_{11} = probability of staying at constant-speed,

π_{12} = probability of transitioning constant-speed to constant-acceleration,

π_{22} = probability of remaining at constant-acceleration,

π_{23} = probability of transitioning constant-acceleration to a 3-G turn,

π_{33} = probability of staying in the 3-G turn,

π_{34} = probability of transitioning 3-G turn to 3-G climb or dive,

π_{43} = probability of transitioning from 3-G climb to dive to 3-G turn,

π_{44} = probability of remaining in a 3-G climb or dive.

5.7 TRACKING BALLISTIC MISSILE TARGETS

The basic ballistic missile tracking problem is illustrated in Figure 5.7.

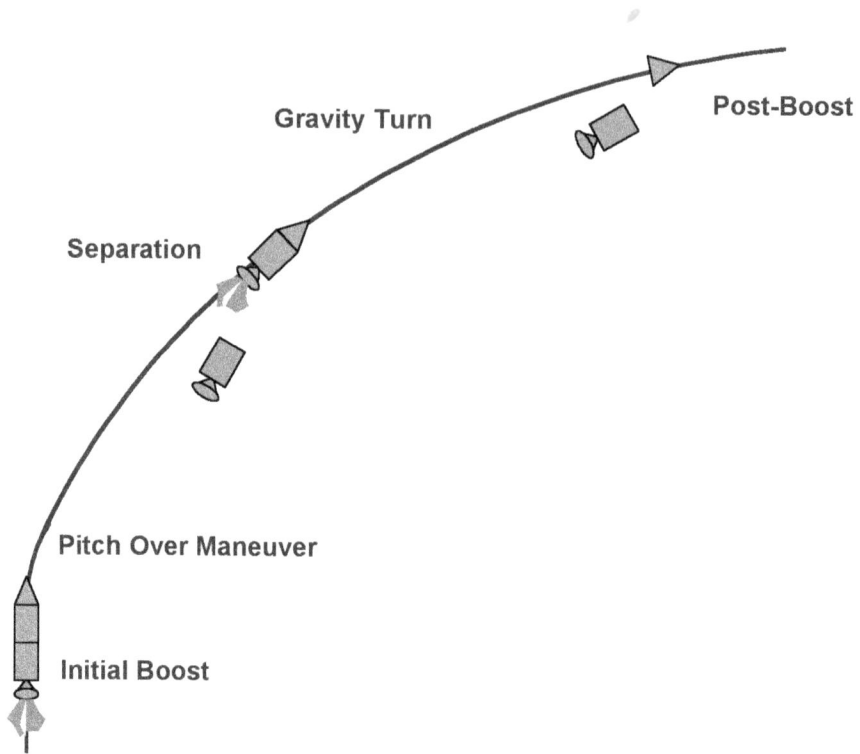

Figure 5.7   Basic Ballistic Missile Initial Trajectory

As can be seen, there are five phases during early flight:

• Initial rocket burn

• Pitch over maneuver

• Object separations

• Gravity turn

• Boost cessation.

After the missile reaches apogee and enters the descent portion of its flight, additional phases of flight are:

• Ballistic flight under force of gravity

• Slow-down due to drag when within the earth's atmosphere.

Use of an IMM tracking filter for ballistic missile targets is illustrated in Figure 5.8. The transition probabilities used in the IMM filter allow a well-organized method of adding a priori knowledge of the target trajectory characteristics to be incorporated into the tracking system. As missiles transition from boost to ballistic, ballistic to reentry, and eventually transition from reentry to a possible ma-

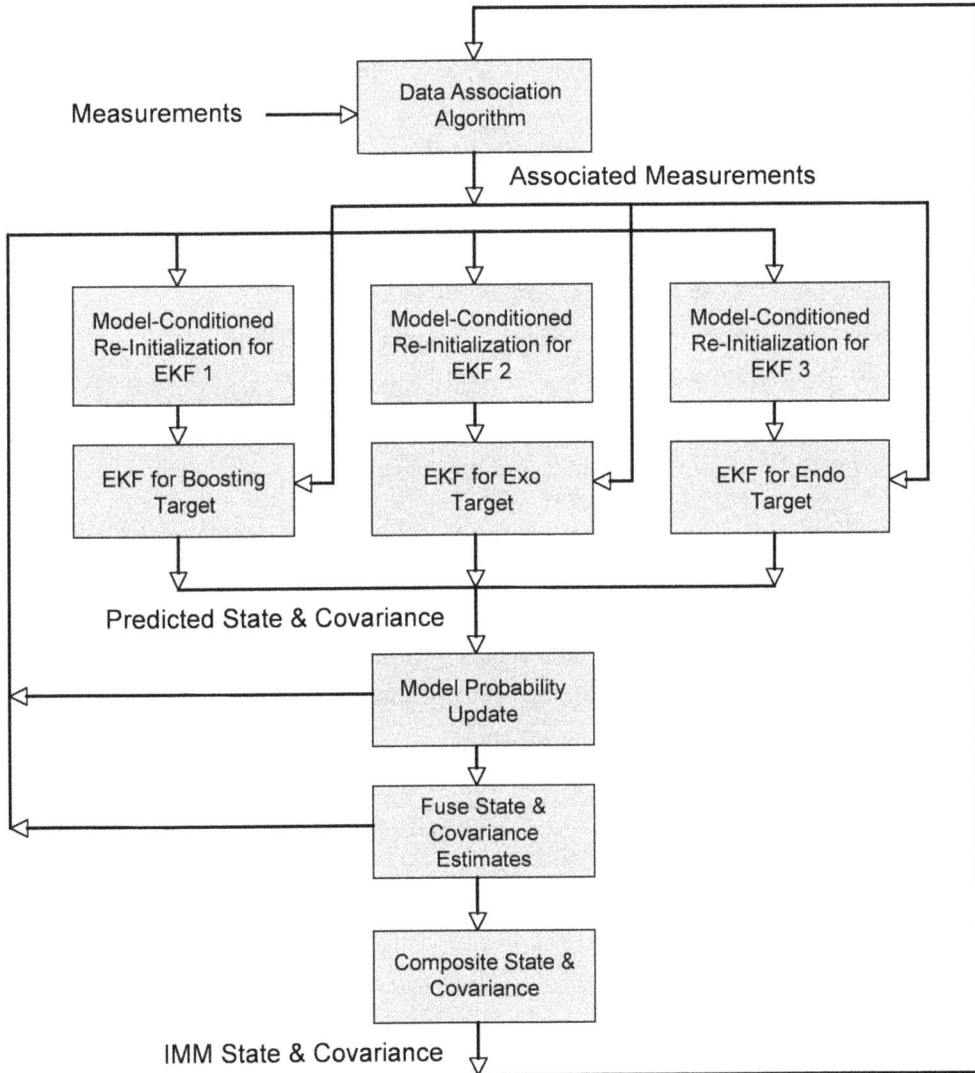

Figure 5.8 Example of Parallel Filters in IMM Approach

neuver and back to reentry a number of distinct state transitions can be defined. The associated dynamical models are described in Table 5.2. Possible target state transitions are given in the following transition matrix.

$$\Pi = \begin{bmatrix} \pi_{11} & \pi_{12} & 0 & 0 \\ 0 & \pi_{22} & \pi_{23} & 0 \\ 0 & 0 & \pi_{33} & \pi_{34} \\ 0 & 0 & \pi_{43} & \pi_{44} \end{bmatrix}, \text{where}$$

π_{11} = probability of remaining in the boost phase,

π_{12} = probability of transitioning from boost to ballistic,

π_{22} = probability of remaining in the ballistic phase,

π_{23} = probability of transitioning from ballistic to reentry,

π_{33} = probability of remaining in the reentry phase,

π_{34} = probability of transitioning from reentry to maneuver,

π_{43} = probability of transitioning from maneuver to reentry,

π_{44} = probability of remaining in the maneuver phase.

Figure 5.9 illustrates the track transitions with a state transition diagram. The environment in which ballistic missile target tracking occurs is much more challenging than for air target tracking due to the large number of possible closely spaced objects associated with the missile complex under track. Figure 5.10 illustrates a simplified view of the source multiple objects to be tracked in this environment.

Table 5.2   Ballistic Missile Dynamics Models versus Trajectory Phase

Object Type	Dynamics Models	Key Required Parameters
Boosting object within Earth's atmosphere	Acceleration, drag, and gravity are modeled	Missile acceleration during boost, drag parameters, and missile mass
Ballistic object within Earth's atmosphere	Drag and gravity are modeled	Missile drag parameters, mas, and velocity at booster burnout (V_{bo})
Ballistic object outside Earth's atmosphere	Gravity is modeled	Missile mass, and velocity at exit of Earth's atmosphere

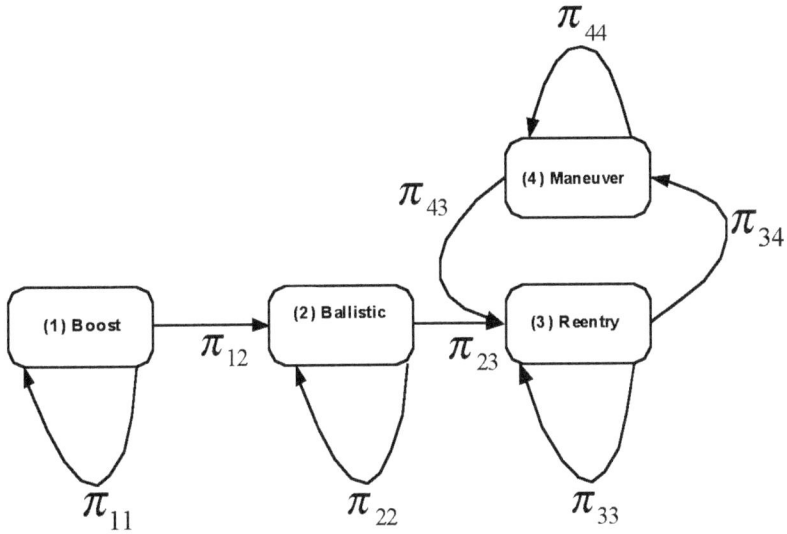

Figure 5.9   Possible State Transition Diagram for Ballistic Missile Targets

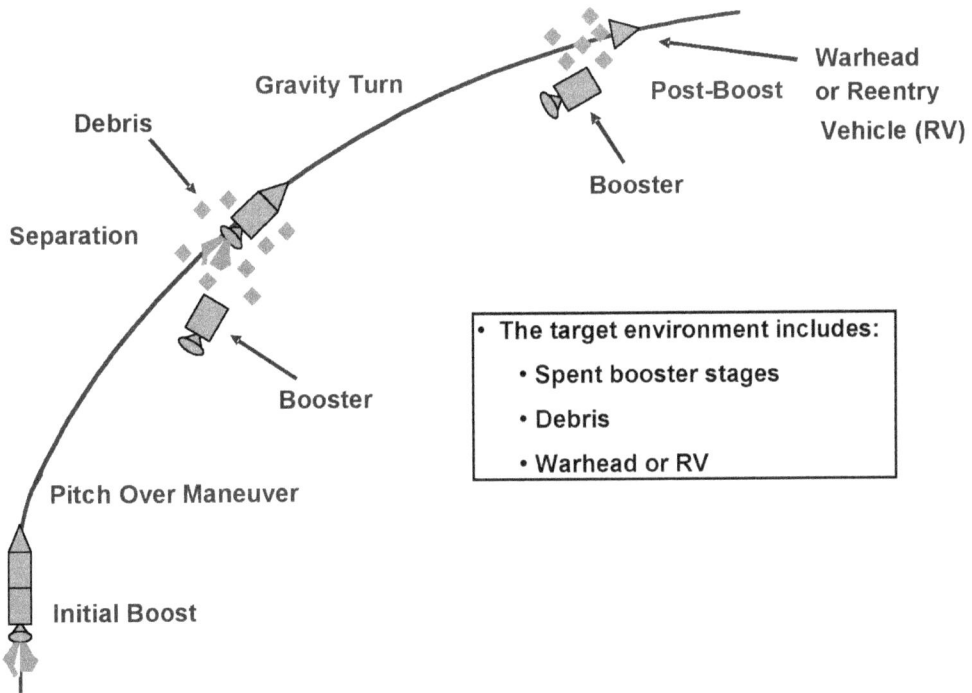

Figure 5.10   Simplified Ballistic Missile Tracking Environment

5.8 TRACKING SURFACE TARGETS

Surface targets are the least challenging from a target-dynamics perspective. However, since they are typically immersed in sea clutter, consistent detection of ship targets becomes problematic. Also, the occurrence of detections due to residual clutter after Doppler processing via either MTI or pulse-Doppler processing may be "spiky" or noise false alarm-like in nature.

Due to the possibility of slowly moving real target and clutter returns in close proximity (and perhaps unresolved in range), the data association problem is more challenging for ship-tracking, especially in high sea-state conditions. Use of MHT techniques is highly recommended to mitigate loss of slow-moving ships in heavy sea clutter, as well as to possibly implement track-before-detect methods when signal-to-clutter ratios are marginal after Doppler processing.

Due to the slow target dynamics involved for ship tracking, relatively low-rate track-while-scan techniques can usually be employed with fixed-gain or simplified Kalman filters, as opposed to the much higher data rates and more complex tracking filters required for air and ballistic missile tracking.

5.9 REFERENCES

[1] Y. Bar-Shalom, *Multitarget/Multisensor Tracking: Applications and Advances,* Artech House, 2000

[2] Y. Bar-Shalom & X. Li, *Multitarget-Multisensor Tracking,* YBS, 1995

[3] S. Blackman & R. Popoli, *Design and Analysis of Modern Tracking Systems,* Artech House, 1999

[4] E. Brookner, *Tracking and Kalman Filtering Made Easy,* Wiley-Interscience, 1998

[5] A. Gelb, *Applied Optimal Estimation,* MIT Press, 1974

[6] S. Haykin & A. Steinhardt, *Adaptive Radar Detection and Estimation,* Wiley-Interscience, 1992

[7] H. Van Trees, *Detection, Estimation and Modulation Theory,* Part 1, Wiley-Interscience, 2001

[8] F. Daum, "A System Approach to Multiple Target Tracking," Chapter 6 in *Multi-Target Multi-Sensor Tracking,* Volume II, edited by Yaakov Bar-Shalom, Artech House 1992

[9] W. Koch & G. van Keuk, "Multiple Hypothesis Track Maintenance with Possibly Unresolved Measurements," *IEEE Transactions on Aerospace & Electronic Systems,* Vol. 33, pages 883–892, 1997

[10] F. Daum, "Book Review: Multiple Target Multisensor Tracking," *IEEE AES Systems,* September 1996

6

Target Classification, Discrimination, and Identification

6.1 INTRODUCTION

This chapter covers the concepts of target classification, discrimination, and identification. [2, 3, 5] are excellent sources of background on this subject. The topics covered in this chapter include:

• Introduction to the target classification problem

• Radar-measured target features

• Waveforms and signal processing

• Feature extraction

• Classifiers:

- Bayes'
- Dempster-Shafer
- Decision trees
- Others

- Classification of air targets:

  - Noncooperative target recognition
  - Target identification (ID)

- Classification of ballistic missile targets

  - Discrimination

- Hit or kill assessment.

The target classification, discrimination, and identification topic completes the overview of fundamental radar theory that forms the basis of designing and analyzing phased-array radars. As noted in Chapter 5, the concept of parameter estimation is at the core of target classification. Here it is referred to by the special name of target feature extraction.

This chapter focuses on the target classification problem that exists for air and missile targets. As will be seen, the detection and tracking of targets is a prerequisite for performing this function. In the case of air targets as described in the previous chapter, the terms *target classification* and *non-cooperative target recognition* (NCTR) are used synonymously. Another term, *identification* (ID), is used as a refined type of classification or NCTR. Although not discussed explicitly in the chapter, ship target classification is very similar to that of the air target case.

For ballistic missile targets, the terms *classification* and *discrimination* are frequently used ambiguously and inconsistently. In this book, *target classification* means to categorize targets by class, such as tactical ballistic missiles (TBMs), intercontinental ballistic missiles (ICBM), intermediate-range ballistic missiles (IRBMs), and so on. Discrimination on the other hand refines classification to object types. A term used in this book for the complete set of possible categorization is *classification, discrimination, and identification* (CDI).

The last part of the chapter addresses the topic of hit or kill assessment. This is included for the reason that air and ballistic missile defense fire control radars usually need to assess the success of the threat intercept when there is adequate time to take a second shot if the first attempt misses. This function is very similar to the target classification problem, using its own unique features to decide on a hit, kill, or miss.

6.2 THE TARGET CLASSIFICATION PROBLEM

In its simplest form, the target classification problem asks the question: What kind of target is being tracked? Since the decision will be based on data or features collected by the radar, it is best expressed mathematically:

$$find\ i\ such\ that\ \left\{\ p(H_i\,|\,\vec{f}\,)\ \right\}\ is\ \max imum\ \geq\ p_{min},\qquad (6.1)$$

where H_i, $\vec{f}$, and p_{min} are the ith target class hypothesis, the target feature vector, and the minimum desired probability of declaring a target class. The conditional probability in equation (6.1) is referred to as the a posteriori or posterior probability, that is, the likelihood that the target is in class i given that feature vector f was measured by the radar.

The test against a minimum probability is optional; however, it is a good practice to apply this type of test to ensure that only reasonably probable class declarations are accepted. In many applications, the minimum probability is supplied to the radar or performed by the command, control, battle manager (C2BMC) or ship combat system controlling the fire control system.

The remainder of this chapter addresses the target features, the radar waveforms to collect them, and the classifiers used to implement equation (6.1).

6.3 RADAR-MEASURED TARGET FEATURES

The feature vector f in equation (6.1) represents the set of all target features collected to perform the target classification function. Possible features include:

• Kinematics (i.e., track-based features)

• Signatures

• Pattern-based.

The first two are physics-based features. Possible target kinematics features include:

• Speed

• Acceleration or deceleration

• Altitude and altitude-rate.

Similarly, signature features can consist of:

• Radar cross section (RCS)

• Target size

• Target shape

• Phase measurements.

Pattern-based features are descriptive of the distribution of objects. At the macroscopic level, an example is targets in a certain formation.

All three classes of target features are useful in classifying, discriminating, and identifying air, missile, and ship targets.

6.4   WAVEFORMS AND SIGNAL PROCESSING

6.4.1   Classification, Discrimination, and Identification Waveforms

Target features are collected by the radar to perform the CDI functions. The waveforms used to enable feature measurements vary with the desired type of features to be collected. The kinematics features listed in Section 3 are usually available from the waveforms used for tracking targets. In general, these are relatively narrowband waveforms. Since most target tracking (with the exception of the track-while-scan approach) use update rates of 1 Hz or higher, in normal operation no additional waveforms need to be scheduled for CDI purposes to collect kinematics features. For low-altitude operation, moving target indicator (MTI) or pulse-Doppler waveforms may be required for detection and tracking. In these cases, kinematics-type features can be extracted from the pulse train.

For signature features, a wide range of waveform bandwidths can be employed, from narrowband to wide bandwidths. Again, where multipulse waveforms are used for clutter mitigation or to measure range-rate, features are extracted from the pulse train returns.

6.4.2   Signal Processing

In the case of narrowband waveforms, such as those used for target tracking, no special signal processing is required, whether single or multiple-pulse waveforms are employed. For the former waveforms, typical signal processing will consist of all-range digital pulse compression (for linear-frequency modulation [FM] waveforms), followed by range and amplitude interpolation and peak-detection. In the latter situation, when multiple-pulse waveforms are employed, pulse matched filtering will be followed by Doppler processing and the above post-detection sequence.

Wideband waveform processing is dependent on the bandwidths used for feature collection. For bandwidths less than 100 MHz or so, current analog-to-digital converter (A/D) technology allows digital pulse compression. However, at bandwidths above 100 MHz, some form of "stretch" or spectrum analysis-type processing will usually be required for matched filtering. For wideband multipulse waveforms, these pulse matched filters will be followed by Doppler processing.

Again, range and amplitude interpolation and peak detection are necessary for wideband waveforms, as well as fine phase measurement for certain feature extraction purposes.

6.5 FEATURE EXTRACTION

The term *feature extraction,* as used in this chapter, covers a broad family of radar measurement processing. For the features described in Section 3, possible feature extraction might entail:

- Standard track filter processing for:
  - Target speed and acceleration
  - Target altitude and altitude-rate (which may require conversion of state vector data)
  - Target rotation-rate and acceleration (depending on state vector composition)
- Computation and smoothing of target RCS
- Computation and smoothing of target size
- Computation and smoothing of fine phase measurements.

Since the tracking filter performs smoothing as a part of its normal processing, no additional smoothing is required for the kinematics feature extraction listed above.

6.6 CLASSIFIERS

As stated in other parts of this book target classifiers can be categorized as being Bayesian or non-Bayesian. In other words, either formal Bayes' rule-type classifiers are employed or those that use other means to decide on target class. The latter category of classifier can be probability-based or not, depending on the specific decision processing implemented.

6.6.1 Bayes' Classifier

The Bayes classifier is an implementation of Bayes' rule of conditional probability [2, 3]:

$$P\left(c_j\middle|f_i\right) = \frac{P\left(f_i\middle|c_j\right)P\left(c_j\right)}{\sum\limits_{k=1}^{M}P\left(f_i\middle|c_k\right)P\left(c_k\right)},\tag{6.2}$$

where $P(c_j|f_i)$ is the probability of target class j given that feature i is measured, $P(f_i|c_j)$ is the conditional probability of feature i occurring given that c_j is the underlying target class j, and $P(c_j)$ is the class prior probability (i.e., the probability of class j occurring out of all J classes).

The two conditional probabilities in equation (6.2) are also referred to as a posteriori (or posterior) and feature probabilities. The J posterior probabilities are the classifier's outputs, and the feature means and class probabilities are elements of the classifier database. The feature probabilities are computed based on the underlying probability density, the feature means, and the error covariance matrix, defined as feature mean value:

$$\mu_{ij} = E\left\{f_i\middle|c_j\right\},\tag{6.3}$$

where f_i and c_j are the ith and j^{th} feature and target class, respectively, and feature error covariance matrix M:

$$M = E\left\{\tilde{f}\,\tilde{f}^T\right\} = \begin{bmatrix} \sigma_{11}^2 & \rho_{12}\,\sigma_1\,\sigma_2 & \cdots \\ \rho_{12}\,\sigma_1\sigma_2 & \sigma_{22}^2 & \cdots \\ & & \\ \rho_{1N}\,\sigma_1\,\sigma_N & \cdots & \sigma_{NN}^2 \end{bmatrix} \tag{6.4}$$

to result in feature probability:

$$P\left(\vec{f}\,\middle|\,c_j\right) = \frac{1}{\left(\sqrt{2\pi}\right)^N |M|^{\frac{1}{2}}}\; e^{-\frac{\left(\vec{f}-\vec{\mu}\right)^T M^{-1}\left(\vec{f}-\vec{\mu}\right)}{2}}. \tag{6.5}$$

In equation (6.5), $\vec{f}$ is the measured feature vector, and $\vec{\mu}$ is the feature mean vector, and M is the feature error covariance matrix. When all features are independent and uncorrelated, equation (6.5) can be simplified to:

$$P\left(f_i\middle|c_j\right) = \frac{1}{\sqrt{2\pi}\,\sigma_{ij}}\; e^{-\frac{\left(f_i-\mu_{ij}\right)^2}{2\sigma_{ij}^2}}. \tag{6.6}$$

In real-world systems, a battle manager, command and control, or combat system will establish a minimum threshold test for the posterior probabilities to declare a target class. Equation (6.2) can be implemented recursively, where posterior probabilities can be used as prior probabilities on successive iterations. When posterior probabilities do not clearly indicate a single-class decision, the battle manager or combat system can defer its decision.

One necessary requirement of a Bayes' classifier is that all possible target classes must be identified in the classifier database. This is required since the Bayes' classifier will always compute posterior probabilities even if the correct class is not one of the target hypotheses (and at least one posterior probability will always be the largest). This is the reason for the minimum probability test implied in equation (6.1). An incomplete classifier database (i.e., with unrepresented target hypotheses) can lead to spurious and erroneous results when only the largest posterior probability is used as the metric to declare target classes. One solution for this inherent problem with the Bayes' classifier is to define an unknown or "strange" class to accommodate nonidentified target classes. When this method is used, the strange class posterior probability can be used to assess the reasonableness of the apparent target class indicated by other posterior probabilities. Such an approach is very important in effectively using Bayes' classifiers and is analogous to adding process noise to a Kalman filter to compensate for unmodeled target dynamics or states.

Given this limitation, the Bayes' classifier using equation (6.5) is the optimal linear classifier when assumed feature probability distributions match the true underlying statistics. When underlying probability densities are known a priori or can be estimated from measurements, these can be optimally used by the Bayes' classifier.

6.6.2   Dempster-Shafer (D-S) Classifier

The Dempster-Shafer (D-S) classifier is a non-Bayesian statistical classifier that uses the concepts of "evidence," "plausibility," and "probability masses" upon which to base target class decisions.

Analogous to the conditional probability used by the Bayes' classifier, the conditional probability mass of class A given features $v1$ and $v2$ can be expressed as:

$$m(A\,|\,v1,v2) = [m(A\,|\,v1)m(A\,|\,v2) + m(AvB\,|\,v1)m(A\,|\,2) + m(A\,|\,v1)m(AvB\,|\,v2)]/D$$

$$(6.7)$$

The probability mass for $m(B\,|\,v1,v2)$ can be expressed in a similar fashion as in equation (6.7). Now consider the probability mass associated with classes A or B conditioned on the features:

$$m(AvB\,|\,v1,v2) = [m(AvB\,|\,v1)\,m(AvB\,|\,v2)]/D \qquad (6.8)$$

where D equals the sum of the numerators and AvB means class A or B.

After all evidence has been considered, the D-S classifier needs a decision rule such as the plausibility of A given by:

$$P(A) = [m(A) + m(AvB)] \,/\, [m(A) + m(AvB) + m(B) + m(AvB)]. \qquad (6.9)$$

As described in [1, 6], the evidence leading to a decision is the probability masses associated with the candidate target hypotheses. Using the mass combination rules, such as that represented by equation (6.9), the plausibility of the underlying target classes can be computed.

Key differences between D-S and Bayes' are the use of unnormalized probabilities (i.e., the probability "masses"), and a probability distribution-free approach compared with the Bayes' classifier, which often assumes an underlying Gaussian probability distribution. Another important difference is the ability to handle correlated features. Bayes' theory incorporates feature correlation information via the feature error covariance matrix, and specifically by the off-diagonal terms. D-S theory does not account for feature interdependencies. For radar applications, this can be a deficiency of the D-S classifier compared with the Bayes' methods. Although the D-S classifier can be modified to account for correlated features, these adjustments are ad hoc in nature and are suboptimal solutions compared to the Bayes' classifier. For this reason, the use of Bayes classifiers for radar-based CDI is often the prevalent choice.

6.6.3 Decision Tree Classifiers

One of the simpler targets classifiers is a decision tree with fixed structure and decision rules. Decision trees are desirable when minimizing computer throughput is a strong consideration in classifier selection and feature statistics are not available or cannot be quantified. Decision trees can employ nonquantitative features and concepts such as "slow targets" versus "fast targets," or "short targets" versus "long targets," "manned targets" versus "unmanned targets," and similar "fuzzy" target-related attributes.

A key rule in designing decision trees is to employ the highest quality features or those with the greatest discriminating capabilities early in the decision process, and lower quality or less discriminating features later in the tree. Figure 6.1 depicts a simple decision tree for use of target total energy (i.e., potential plus kinetic energies) to separate tactical ballistic missiles and air-breathing targets (ABTs).

6.6.4 Rule-Based Classifiers

Rule-based algorithms can be used for target classifiers These may or may not use quantitative features and can make "hard" or "soft" decisions, unlike simple

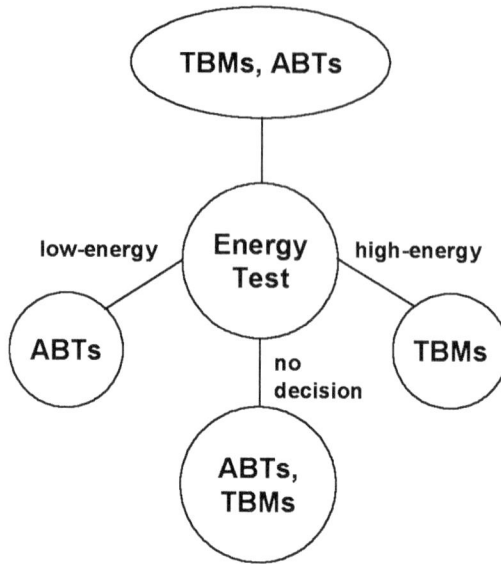

Figure 6.1   A Typical Decision Tree Classifier

decision trees that only make "hard" decisions (e.g., a target is in class A or B not, perhaps, in both classes). These rules are usually logical functions such as "if-then-else." An example of rule-based classifier constructs are:

> If {speed is slow}
>
> Then{targetisahelicopterorunmannedaerialvehicle(UAV)}(6.10)
>
> Else {target is a tank or ship}

or,

> If {speed > v_p}
>
> Then {target is jet-powered}   (6.11)
>
> Else {target is propeller powered}.

As can be seen, either qualitative or quantitative rules can be used in a decision tree. This also provides the ability to use so-called fuzzy logic or neural-like processes. The primary drawback of these classifiers is from an algorithm-training and analysis perspective, since there is no systematic analytical method of designing or analyzing them.

6.6.5 Compound Classifiers

The classifiers discussed in the preceding sections are some commonly used ones; there are many more defined in [2, 3]. Another possible classifier type is based on the combination of one or more of these (and other) algorithms to create a "compound" classifier.

Often, target classification processing might use a decision tree at a high level as the overall classifier structure. Each node can then employ different types of classifiers such as Bayes', D-S, or rule-based approaches. Since each of the classifiers discussed in this chapter has strengths and weaknesses, the best solution is to select the best classifier for separating or classifying targets at the particular stage of classification processing based on performance, efficiency, and so on. This approach can yield a very powerful solution technique for target classification, discrimination, and identification problems encountered in radar applications.

6.7 CLASSIFICATION OF AIR TARGETS

Air target classification can use any or all of the classifier types described in Section 6. The desired end result is to categorize the tracked targets into distinct classes or types so that subsequent radar processing can be performed, such as interceptor support in the case of fire control systems.

In addition to the kinematics, signature, and context-based features described in Section 3, additional data are often available for classifying air targets. One type of data is Identification, Friend or Foe (IFF). This data is available from cooperative targets that use an IFF transponder. This allows easier classification of friendly aircraft and ships that use IFF transponders. Another specific context-like feature is procedural in nature. Operation rules such as flight corridors can be defined to control and identify friendly aircraft by requiring that they fly within these corridors (an exception being the case of damaged aircraft that have limited flight capability or where safety over-rules the use of corridors). These types of techniques along with IFF represent powerful target ID features.

Air-breathing targets exhibit a number of specific kinematics and signature features, including:

• Speed, acceleration, altitude, and altitude-rate
• Observed maneuver capability

- RCS
- Estimated size
- Target shape.

   The above features can be used with the Bayesian and non-Bayesian classifiers described in Section 6 to decide on the likely target class, such as:

- Aircraft
- UAV
- Helicopter
- Cruise missile
- Other.

   Moreover, these techniques can also be used to refine categories to types, or to perform the ID function, such as airframe type.

   The target classes and types listed above, along with their associated posterior probabilities (when using Bayes-type classifiers) can be provided to the fire control system.

6.8 CLASSIFICATION OF BALLISTIC MISSILE TARGETS

The classification of ballistic missiles (BMs) is much different from that for air targets. Although some similar target features are employed, their values and specific usage differs. BM targets can exhibit features such as:

- Speed, acceleration, altitude, and altitude-rate
- Observed maneuver capability
- RCS
- Size.

   These and other features can be used, preferably by the Bayesian classifier described in Section 6, to decide on the likely target class, such as:

- Theater or tactical BMs
- Intermediate-range BMs
- Intercontinental BMs.

Discrimination techniques can then be used to further refine categories to types.

The target classes and types listed above, along with their associated posterior probabilities (when using Bayes-type classifiers) are provided to the C2BMC or ship combat system for use in computing intercept solutions.

6.9   HIT OR KILL ASSESSMENT

For systems that allow a shoot-look-shoot firing doctrine when battle space and timeline permit, hit or kill assessment is a valuable radar function. Successful determination of the effectiveness of an intercept can avoid wasting expensive interceptors, or can improve the probability of negation by allocating additional interceptors when available and feasible.

Hit or kill assessment (KA) is much the same as target classification, except that here the classes of interest are:

• Hit

• Kill

• Miss.

Like the target classification problem, hit or kill assessment can employ any of the classifiers described in Section 6.

6.10   PERFORMANCE PREDICTION

Back-of-the-envelope calculations of classification performance are valuable to validate correct operation of a target classifier. One method used for estimating classification performance is the K-factor, defined as:

$$K = \frac{\mu_2 - \mu_1}{\sqrt{\frac{1}{2}\left(\sigma_1^2 + \sigma_2^2\right)}},\qquad(6.12)$$

where the μ_1, μ_2, σ_1^2, and σ_2^2 are the mean values and variances of feature 1 and feature 2, respectively. Since the K-factor is a normalized statistical distance, if the underlying probability densities for the feature distributions are Gaussian, then the probability of correct classification can easily be calculated using the appropriate K-factor from either equation (6.12), or (6.13) as described in the following paragraph.

When multiple statistically independent features are used by a classifier, an aggregate K-factor can be calculated:

$$K_{TOTAL} = \sqrt{K_1^2 + K_2^2 + K_3^2 + \cdots + K_M^2} \ , \tag{6.13}$$

where through are the individual K-factors for the M features, calculated using equation (6.12).

6.11 REFERENCES

[1] P. Dempster, et al., *Classic Works on the Dempster-Shafer Theory of Belief Functions*, Springer, 2007
[2] R. Duda, et al., *Pattern Classification*, 2nd Edition, Wiley-Interscience, 2000
[3] K. Fukunaga, *Introduction to Statistical Pattern Recognition*, 2nd Edition, Academic Press, 1990
[4] A. Gelb, *Applied Optimal Estimation*, MIT Press, 1974
[5] S. Theodoridis & K. Koutroumbas, *Pattern Recognition*, 2nd Edition, Academic Press, 2003
[6] G. Shafer, *A Mathematical Theory of Evidence*, Princeton University Press, 1976

7

Data Processing Algorithms for Phased-Array Radars

7.1 INTRODUCTION

This chapter covers a variety of radar data processing algorithms used by phased-array radars (PARs). These algorithms are usually implemented in software, many of them in a mission application computer program and others in areas such as signal processing and alignment and calibration software. Unlike the previous chapters that treat topics covered in many theory and application texts, these topics are not usually discussed in radar systems books. They are often covered in technical journal papers and related literature, but not in a unified treatment. The objective of this chapter is to collect these data processing algorithms in one place, organized by algorithm purpose and type.

Algorithms discussed include:

- Resource management (RM) and planning
- Radar waveform scheduling
- Search functions
- Target data association

- Statistical tracking filters
- Target feature extraction
- Classification and discrimination
- Radar hardware commands generation
- Returns processing
- Waveform matched filtering
- Detection processing
- Monopulse processing
- Coherent and non-coherent integration
- Pilot pulse calibration
- Satellite calibration
- Digital beam forming (DBF)
- Sidelobe cancellers (SLCs)
- Adaptive processing
- Statistical signal processing.

These represent commonly required algorithms necessary for many radar applications.

7.2   DATA AND SIGNAL PROCESSING ALGORITHMS

The previously listed algorithms are grouped into the following categories:

- Resource planning and radar scheduling
- Search and track
- Classification, discrimination, and identification (CDI)
- Radar hardware control
- Radar measurement processing
- Signal processing
- Calibration and alignment

- Adaptive processing

- Statistical estimation and detection.

The following sections describe these categories of algorithms.

7.2.1   Resource Planning and Radar Scheduling Algorithms

7.2.1.1 Resource Management. All phased-array radars (PARs) use some form of resource management algorithms for the purpose of allocating radar duty factor and timeline occupancy to the activities that the radar will execute. Many solid-state PARs are duty-limited to a typical range of 20 to 30 percent over a short duration. For multifunction radars (MFRs) this duty factor is a resource to be assigned or allocated to specific functions such as search, tracking, and CDI.

In addition to duty factor, the radar timeline must also be managed, as in many cases it is more of a limited resource than the transmit duty. Timeline occupancy; that is, the total of all transmit and receive actions executed by a radar— must be kept to no greater than 100%. In general, search activities with large range extents, and equivalently long duration receive windows, use the most timeline of all radar functions. Since many MFRs perform search about 30 to 50% of the time, search is often the driving factor for timeline occupancy usage.

More specifically, RM algorithms are typically based on relative radar function and activity priorities. These priorities can be static (i.e., fixed) or dynamically changing depending on the particular mission application the radar is performing.

Many PARs manage duty and occupancy allocations via use of some form of radar activity priority (RAP) tables. Again, these tables can be static assignments of a priority or rank to each function the radar can perform, or they can be dynamically changing as a function of the radar's environment, loading, or other conditions. Often these RAP tables are static, while individual activities within an activity class have dynamic rank or priority.

An example of a RAP table appears in Table 7.1. The RAP table indicates that search is the highest-priority radar task, whereas diagnostics is the lowest-priority function performed by the radar.

Table 7.1 Example of a Radar Activity Priority (RAP) Table

Radar Activity	Relative Rank or Priority
Search	Highest
Track initiation	
Track maintenance	
Cued search	
CDI	
Re-acquisition	
Pilot pulse calibration	
Diagnostics	Lowest

Many RM algorithms are hierarchical in nature, often consisting of longer- and shorter-term planning. For example, the long-term planner (LTP) might allocate radar activities to short-term planning (STP) intervals, leaving specific assignment to resource intervals (RIs) or resource periods (RPs) to the STP or radar scheduling algorithm.

Specific implementation of RM algorithms varies with radar mission. Some mechanically steered PARs use long LTP intervals to accommodate the inertia of moving antenna pedestals or antenna mounts. Other radars might eliminate LTPs (and even STPs) completely due to very short reaction timelines, for example, for ship self-defense (SSD) in the presence of high-speed incoming missiles. Most air and ballistic missile defense radars use some form of LTP/STP to manage the radar's resources for activities that can be planned in advance (e.g., tracking at a fixed update rate).

For this reason, LTP intervals can vary from as long as 300 seconds to as short as 1 second or less, while STP intervals can vary from a few seconds to fractions of a second. Usually the STP interval is selected to match the long-term scheduler (LTS) interval, which is described in the next section.

7.2.1.2 Radar Scheduler. The radar scheduler (RS) algorithms assign radar transmit and receive actions to the timeline as allocated by the previous STP function. A radar scheduler is usually implemented using two sub-functions: the long-term scheduler and the short-term scheduler (STS). Radar schedulers are typically structured in this hierarchical form for several reasons, where the primary differences between the LTS and STS are:

- LTS is usually matched to the duration of the STP interval

- STS is usually matched to the scheduling interval (SI) or resource interval/period (RI/P)

- LTS translates preplanned STP activities to transmit and receive actions

- STS adds dynamically occurring transmit and receive actions to the radar timeline (e.g., search verify or track initiation [TI] waveforms) that cannot be preplanned.

Three types of scheduler are described in the following subsection. One form of the hierarchical RM/RS structure is shown in Figure 7.1.

Single-Function Scheduler.   This type of scheduler is typically dedicated to a single radar function or mission. One example is target tracking for a ship self-defense application. Since this function's purpose is the detection and tracking of incoming, high-speed missiles, its timelines are extremely short (e.g., typically matched to the coherent waveform dwell time). For cases such as this, there can be insufficient time to use the hierarchical RM and RS structure shown in Figure 7.1. Instead, each pulse or pulse train is scheduled using a priori target acquisition and tracking strategies to simplify and limit the required RS processing. In fact, the primary input to this type of RS algorithm is usually the tracking data. These data are used to schedule subsequent track maintenance (TM) waveforms.

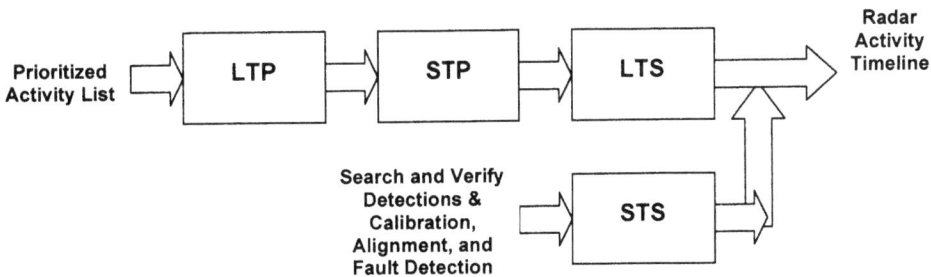

Figure 7.1   Basic Hierarchical RM/RS Structure

Multifunction Schedulers.   There are many types of multifunction schedulers, used for example in early warning and air and missile defense fire control applications. Two major classes of multifunction RS are template-based and adaptive algorithms. These are discussed in the following subsections.

Template-Based Schedulers.   Template-based schedulers can be implemented in several ways. However, the key attribute of template-based RS is the use of predefined radar activity patterns. Figure 7.2 illustrates an early warning radar waveform template. This fixed template interleaves search (S), tracking (T), and fault-detection (FD) in a repeating pattern.

Adaptive Schedulers.   Again, there are many implementations of adaptive scheduling. The key attribute of an adaptive scheduler is its capability to dynamically schedule radar actions based on the prevailing environment and radar function priorities. Rather than using a fixed pattern, an adaptive scheduler will place transmit and receive actions on the timeline in a priority fashion in a manner that satisfies a number of scheduling rules. These rules for waveform priorities might be, for example, those appearing in Table 7.2.

Hence, based on the rules in Table 7.2, multipulse waveforms, when allocated resources (i.e., planned), are placed on the timeline first, then cued search pulse (when planned) next, and so on. Prior to detecting any objects, only the horizon search will be planned, and therefore it will have the highest scheduling priority.

It should be obvious that this type of scheduler can be more complex than a fixed template-based scheduler. Therefore, this will result in more computationally intensive scheduling algorithms and a larger amount of associated computer code.

Hybrid Schedulers.   When different degrees or levels of flexibility are required of the RS, hybrid scheduling approaches can be used. These hybrid RS algorithms can employ a basic template for assigning functions to resource intervals, such as in Figure 7.1, but use an adaptive approach to place specific waveforms

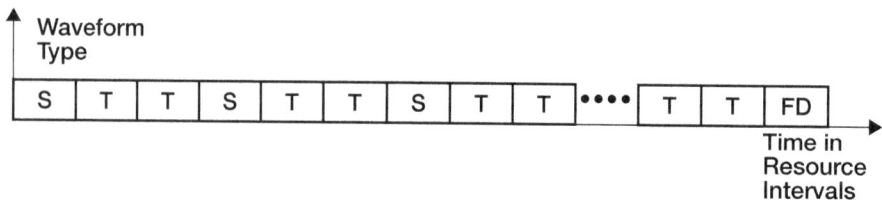

Figure 7.2   Early Warning Radar Scheduling Template

Table 7.2   Example of Adaptive Radar Scheduling Rules

Waveform Type	Scheduling Priority
Coherent multi-pulse	Highest
Non-coherent multi-pulse	
Cued search pulses	
Verify pulses	
Track initiation pulse-pairs	
Track maintenance pulses	
Wideband discrimination pulses	
Horizon search pulses	Lowest

within a given resource interval. In other words, several scheduling intervals could be created within a single resource interval using a rule-based or adaptive algorithm, where a template approach could be used at the resource interval level.

7.2.2   Search and Track Algorithms

The following subsections cover the class of common search and track algorithms employed by PARs.

7.2.2.1 Search and Acquisition Functions. The following subsections describe some commonly used search and target acquisition algorithms.

Volume and Cued Search.   Volume and cued search are two common types of search used by PARs that perform missions such as anti-air warfare (AAW) and handovers for air and missile defense radars. Large volume searches are commonly employed by AAW radars since they are a fundamental function of these types of radars. AAW and search radars in general usually operate at frequencies of S-band and below. The relatively larger antenna beamwidths at these operating bands compared with higher frequency radars for the same size antenna aperture, minimize the number of beams necessary to execute the volume search. For example, for the same-size aperture, a single S-band beam subtends about 7 to 9 X-band antenna beams, as depicted in Figure 7.3 (the ratio is given by the radar wavelength squared).

Volume searches are typically defined by specifying azimuth, elevation, and range extents, as well as frame time and probability of false alarm. The structure of a possible volume search "raster" is illustrated in Figure 7.4.

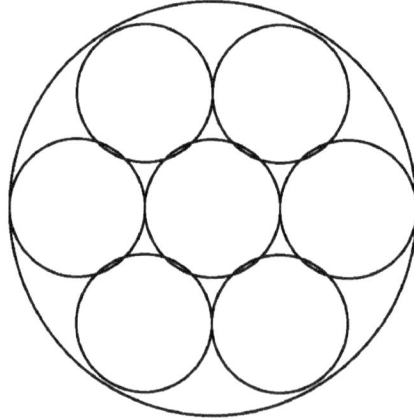

Figure 7.3 Comparison of S-band to X-band Antenna Beams for Equivalent Aperture

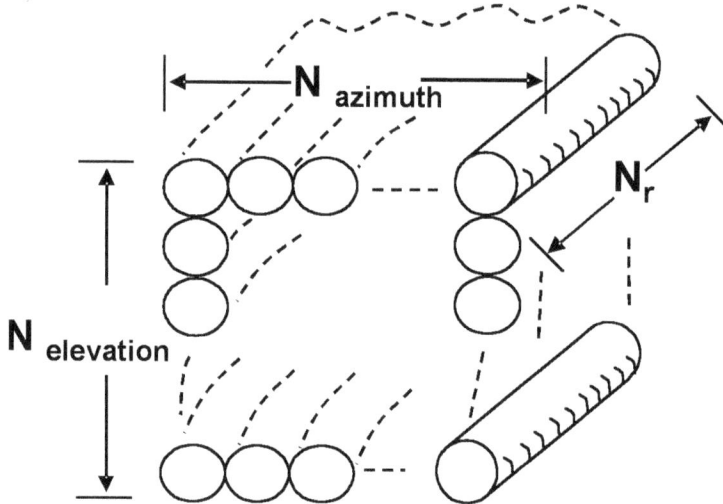

Figure 7.4 Volume Search Raster Structure

Frame time (T) is defined as the time required to complete one complete pass through the search volume. The probability of false alarm for a volume search is typically selected to produce a specified false alarm rate. This minimizes the wasting of radar resources (duty factor and timeline occupancy) in attempting to acquire and initiate tracks on false alarms to an acceptable level. For example, to limit the false alarm rate to 5 per second for a volume search with 100 beams and 1,000 range cells (i.e., 10^5 range-angle cells) to be searched in a frame time of 2 seconds requires a single-cell probability of false alarm of 10^{-4}. This calculation is shown in equation 7.1.

$$P_{fa} = \frac{N_{FA} \cdot T}{N_b \cdot N_r} = \frac{(5)(2)}{10^5} = 10^{-4} \qquad (7.1)$$

For fixed-array radars, beam steering is usually in sine-space coordinates, and angles are expressed using the direction-cosines u and v as opposed to azimuth and elevation for moving-antenna systems. When used for handovers between sensors, the volume search is often centered on a point in inertial space.

Waveforms used for these searches will depend on the environment in which they are used. In the clear, when sufficient single-pulse signal-to-noise ratio (SNR) is available to meet detection requirements, some form of waveform template can be employed. For relatively small range extents where timeline occupancy is not an issue, a template such as that shown in Figure 7.5 can be used. As can be seen, the template consists of a transmit followed by a receive window (equal to the search range extent in duration plus transmit

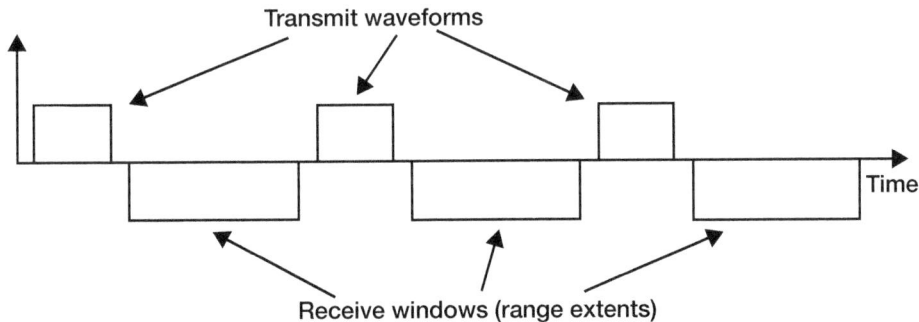

Figure 7.5 Possible Waveform Template for Volume Search

pulse length). However, when many beams are required or the frame time is short, this type of search waveform can exceed its allocated timeline occupancy. Search occupancy can be assessed from the required beam rate (for transmit and receive pairs) to execute the search. The situation where the required beam rate exceeds that available for executing the search is referred to as "occupancy-limited."

For occupancy-limited situations, the waveform template of Figure 7.5 cannot be employed. Instead, a different template must be used. One possible template is shown in Figure 7.6. Here, N sub-pulses are transmitted in sequence (referred to as "shot-gunning") to N different beam positions in the raster (and at different frequencies to avoid cross-talk between adjacent beams). This is followed by N simultaneous receive beams corresponding to the N transmit beams. This approach requires N antenna beamformers (or digital beam forming), receiver channels, and signal processor channels, but reduces the timeline occupancy by nearly a factor of N for very large search range extents.

Cued searches are really a special form of a volume search where the center of the search in angles and range is specified by a propagated target state vector and validity time supplied by a handover source or "cue." The source of the cue or state vector can be, for example, another radar, the same radar (referred to as self-cueing), or an electro-optics (EO) sensor, which can be located on the earth's surface (e.g., land- or sea-based), in the air (e.g., airborne sensor such as an aircraft or unmanned aerial vehicle [UAV]) or in space (e.g., satellite). The types of search raster and waveforms used are identical to those for a fixed volume search as for the examples depicted in Figures 7.4, 7.5, and 7.6.

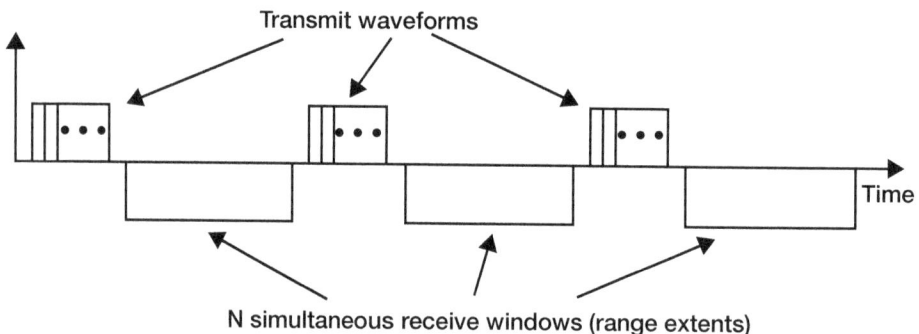

Figure 7.6 Possible Waveform Template for Occupancy-Limited Volume Search

Horizon Search Fences. Horizon searches are a special form of a volume search used primarily by missile early warning and ballistic missile defense radars. Unlike the more general volume search, a horizon search is typically limited to a small number of rows in elevation, often only one. The basic purpose is to detect and acquire ballistic missiles. The premise for these searches are that for sufficiently long-range radars, any approaching and ascending ballistic missile must pass through the horizon fence. These searches are based upon cumulative detection probability, which is discussed earlier in the book.

An example of a basic horizon fence is shown in Figure 7.7. In this example, the fence covers ± 60 degrees in azimuth and is erected at an elevation angle of 3 degrees above the local horizon. In the illustration, the antenna beams are overlapped at 80% of the 3 dB beamwidth. This is a fairly common beam-packing density and provides a nearly flat response in SNR across the azimuth coverage.

The basic fence design strategy is to size the radar using the radar range equation to achieve the desired cumulative probability of detection given that the missile will be in the fence coverage for at least N looks. The probability of false alarm is, like for the volume search case, selected to achieve a desired false alarm rate, per the expression used in equation (7.1).

Search Verification. A key algorithm necessary to preclude attempting to acquire false alarms (and the attendant waste of radar resources) is search verification processing. Although not required to acquire targets, verification is desirable to allow use of higher per-pulse probabilities of false alarm.

The key to this approach is for the radar to return to a beam and range where search detections occurred to corroborate that the detections arose from real targets. Typically, the verify process is to transmit a search-like waveform (e.g., same RF frequency and bandwidth), or a sequence of search-like waveforms, to

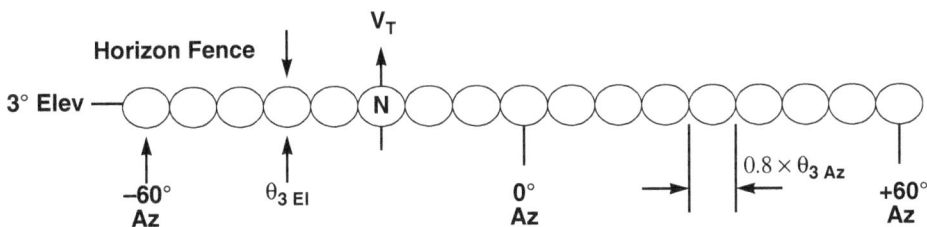

Figure 7.7 Typical Horizon Search Fence

the antenna beam position of the original search detection. When the verify waveform does not corroborate the search detection, the original search detection was in all likelihood a false alarm. However, when the search detection is confirmed by the verify waveform, the presence of a real target is indicated.

Since two waveforms (search and verify) are used, the radar resources and performance probabilities can be allocated across the search and verify steps in the detection process. The balancing between these functions allows optimization of resources (i.e., minimizing the total radar duty and timeline occupancy) to achieve the overall desired probability of false target acquisition. For a false target acquisition to occur when using search and verify requires false alarms on both actions. The probability of this is given by:

$$P_{false\ acq} = P_{fa-search}\ P_{fa-verify} . \tag{7.2}$$

For example, to achieve $P_{fa\ acq}$ of 10^{-6}, the latter two probabilities in equation (7.2) could be 10^{-4} and 10^{-2} for search and verify, respectively.

If a simple Swerling I target fluctuation model is assumed, the probability of acquisition without verification is given by:

$$P_{acq} = \left(P_{false\ acq}\right)^{\frac{1}{1 + SNR_{Acq}}} . \tag{7.3}$$

For the case with verification, the probability of acquisition is given by:

$$P_{acq} = \left(P_{false-search}\right)^{\frac{1}{1 + SNR_S}} \left(P_{false-verify}\right)^{\frac{1}{1 + SNR_V}} . \tag{7.4}$$

Improvement in radar resource utilization is achieved if the energy allocated to the search-plus-verify is less than that required for the search action plus subsequent undesired track initiation actions (i.e., when the search detection is a false alarm), or equivalently when:

$$E_{TI} > E_V . \tag{7.5}$$

Track Initiation.   Track initiation is an algorithm that collects data for the initialization of the target state vector and error covariance matrix used by statistical tracking filters. Since the waveform bandwidth used for tracking is usually different (i.e., greater) than that used for search or verify, the TI function generally uses the same bandwidth as is used for tracking to ensure a better match to the tracking waveform in the transition from TI to track maintenance.

Two common types of TI waveforms are sequences of single pulses and sequences of pulse-pairs. The former approach might employ an M-out-of-N scheme whereby at least M detections are required to establish the target state vector. The latter approach requires detection of both pulses of a given pulse-pair to initiate a track. When linear frequency modulation (LFM) or "chirp" waveforms are used for TI, the latter approach will often employ an up-chirp followed by a down-chirp for the purpose of deriving accurate and unambiguous range and range-rate (i.e., by exploiting the range-Doppler coupling associated with match-filtered LFM waveforms).

7.2.2.2 Target Data Association. The following subsections describe some common algorithms used for performing the data association (DA), that is, returns-to-tracks association, necessary to maintain tracks when using multiple-target tracking filters.

Nearest-Neighbor Algorithms. The nearest-neighbor algorithm is the simplest approach to performing the DA function. The underlying premise of NN is that a detection that is closest to the predicted location of an existing target track is most likely associated with that track. For sparsely populated target environments, the NN algorithm can perform adequately. However, in densely spaced objects or heavy clutter-backscatter situations, NN can perform very poorly. It is, however, the "cheapest" in terms of computer throughput and algorithm complexity.

For radars that measure slant range and two angles (i.e., three-dimensional [3-D] radars), the performance metric is the normalized statistical distance measure of detections and tracks given by:

$$X_{ij}^2 = \frac{\left(R_i - \hat{R}_j\right)^2}{\sigma_R^2 + \sigma_{\hat{R}}^2} + \frac{\left(\theta_i - \hat{\theta}_j\right)^2}{\sigma_\theta^2 + \sigma_{\hat{\theta}}^2} + \frac{\left(\phi_i - \hat{\phi}_j\right)^2}{\sigma_\phi^2 + \sigma_{\hat{\phi}}^2}, \tag{7.6}$$

where each numerator is the difference between a radar measurement and its predicted estimate from the tracking filter, and the error variances in the each denominator are for the measured and estimated quantities, respectively. This distance metric is a random variable, which for Gaussian measurement and estimation errors has a chi-square probability density, which in the case of equation (7.6) is with 3 degrees-of-freedom (DoF). Equation (7.6) can be expanded to incorporate any number of target measurements or features.

The NN approach uses a criterion that associates detection i with track j when:

$$X_{ij}^2 \leq X_{lj}^2 \quad \forall \text{ values of } l .\tag{7.7}$$

Probabilistic and Joint Probabilistic Data Association. The probabilistic DA (PDA) and joint-probabilistic DA (JPDA) algorithms associate probabilities to detection-track pairs for the purpose of association, where the target distribution in the volume is assumed to follow a uniform probability model. JPDA considers multiple tracks and detections jointly rather than independently as is the PDA algorithm. For densely spaced objects or targets in clutter, JPDA can provide superior performance compared with the simpler PDA approach, and vastly superior performance compared with the NN method.

The joint association probability is given by:

$$P\left\{\theta(k)\middle|Z^k\right\} = \frac{1}{c}\frac{\phi!}{m(k)!}\mu_F(\phi)V^{-\phi}\prod_j\left\{f_{tj}\left[z_j(k)\right]\right\}^{\tau_j}\prod_t\left(P_D^t\right)^{\delta_t}\left(1-P_D^t\right)^{-\delta_t}\tag{7.8}$$

where targets are assumed to be uniformly distributed in the volume, and θ_{jt} is the event that measurement j originated from target (i.e., track) t for $j = 1,..., m$, and $t = 0, 1,..., N_T$. The parameter t_j is the index of the track to which measurement t_j is associated in the event under consideration, N_T is the known number of tracks, and V, $\phi(\theta)$, and $\tau_j(\theta)$ are the geometric volume (over which measurements/tracks are assumed to be uniformly distributed), the total number of false unassociated measurements in $\theta(k)$, and the measurement association indicator (as defined in Bar-Shalom [1]), respectively. This technique results in a moderate to high computer throughput usage and a moderately complex data processing implementation. When target spatial distributions are not uniform and are known or can be estimated from measurements, the uniform density assumption can be replaced to yield more optimal performance.

NN-JPDA. The NN-JPDA is a composite algorithm where the NN method described under "Nearest-Neighbor Algorithms" is used as a pre-processor to identify a candidate set of detections that are reasonably "close" to a given track. This step is followed by application of the JPDA algorithm described under "Probabilistic and Joint Probabilistic Data Association which now operates on the subset of detections selected by the NN algorithm. This slight increase in algorithm complexity serves to reduce the computer throughput requirements relative to the JPDA algorithm to a moderate level, and therefore places it somewhere between the NN and JPDA in relative throughput usage.

Multiple-Hypothesis Tracking. Multiple-hypothesis tracking (MHT) is an optimal DA approach that creates and propagates a number of detection-to-track hypotheses to eventually arrive at the correct return-to-track association. In its most general form, hypotheses are formed for each new detection-track pair at each track filter update time. Therefore, the number of possible hypotheses grows geometrically with time.

However, since every possible association is propagated by the MHT algorithm, by definition, all correct return-to-track associations are made (unfortunately, along with a number of incorrect associations). Since each hypothesis is used to update a distinct track filter, usually of the Kalman or extended Kalman filter family, use of MHT can result in progressively increasing computer throughput usage if allowed to continue unabated. Blackman and Popoli [2], describe the basic MHT theory in great detail, so specific algorithm details are not further discussed here.

In real-world applications, reasonably unlikely hypotheses (based on, e.g., physics-based rules) can be eliminated or "pruned" at each stage or track update time. One method of pruning is to propagate and test association probabilities for each hypothesis and eliminate those that do not exceed a specified minimum probability. Another approach is to use a NN-like detection-to-track proximity test to identify hypotheses to be dropped from further consideration in a similar manner to the NN-JPDA approach. These ad hoc pruning techniques must be used judiciously to trade off near-optimal association performance with throughput usage. Like many such algorithms, simulation techniques are usually employed to tune performance for specific applications and target environments.

7.2.2.3 Statistical Tracking Filters. The following subsections describe a number of commonly implemented tracking filters.

Alpha-Beta ($\alpha - \beta$) and Alpha-Beta-Gamma ($\alpha - \beta - \gamma$) Filters. This is the simplest class of tracking filter, both in algorithm complexity and computer throughput usage. Although these filters are not really statistical in nature (i.e., they are deterministically defined), they are covered here since they are still widely used for certain tracking applications. Most filters in this class use "fixed" gains or deterministically selected gain constants to update target state vectors via:

$$\hat{x}_{k+1} = \Phi \hat{x}_k + k_k \left(z_k - H \hat{x}_k \right), \tag{7.9}$$

where k is the constant gain or a deterministically calculated gain from a look-up table.

Equation (7.9) can be rewritten as:

$$\hat{x}_{k+1} = \left(\Phi - k_k H \right) \hat{x}_k + k_k z_k . \tag{7.10}$$

The matrix Φ is the state transition matrix, H is the observation matrix, k is the gain vector, and z_k is the kth measurement vector. When position and rate are filtered, equation (7.10) is referred to as the $\alpha - \beta$ filter. When acceleration is added to the state vector in equation (7.10), the filter is referred to as the $\alpha - \beta - \gamma$ filter.

The calculation of the filter gains can be performed offline using analytical or Monte Carlo methods. The gain can be constant or change periodically via a deterministic function or lookup table. Since there is no real time calculation of gains, these filters are the least complex in software implementation and use the least computer throughput.

Kalman Filters (KFs). These tracking filters are truly in the statistical filter class. The target state vector is modeled as a random process, and the predictor-corrector form of equations (7.9) and (7.10) used for the previous filters are also used by the Kalman filter (KF). However, the gain vector is now a random process that is computed at each filter update. Both the target state vector and error covariance matrix are propagated and updated in the KF algorithm.

The KF gain is computed using:

$$k_k = P_{k+1|k} H \left[H P_{k+1|k} H + R_k \right]^{-1}, \tag{7.11}$$

where $P_{k+1|k}$ and R_k are the predicted state error covariance and measurement covariance matrices, respectively.

The predicted or propagated state error covariance matrix is given by:

$$P_{k+1|k} \;=\; \Phi\, P_{k+1|k}\, \Phi^T + Q_k, \qquad (7.12)$$

where Q_k is referred to as the process noise covariance, which is a measure of the KF model uncertainty (i.e., inaccuracy of the system model embodied by the transition matrix Φ).

A key attribute of the a well-designed or properly tuned KF is the concept of independent and statistically "white" or uncorrelated estimation-error residuals, also referred to as "innovations," where the error residuals are defined as:

$$residual \;=\; z_k \;-\; H\,\hat{x}_k. \qquad (7.13)$$

The statistical whiteness of the residuals refers to the uncorrelated nature of the errors. This is the criterion upon which the KF derivation is based.

The computational driver for KF implementations is the gain computation, equation (7.11), which requires the inversion of an $N \times N$ matrix. There are many numerically efficient and well-conditioned implementations of (7.13) that avoid explicitly performing a matrix inverse.

For linear estimation problems, the KF is the optimal linear filter in the minimum mean-square error (MMSE) sense. For nonlinear estimation problems, the KF is the optimal *linear* filter. Nonlinear solutions are discussed in the next section.

Extended Kalman Filters (EKFs). Extended Kalman filters (EKFs) are linear filters designed to solve inherently nonlinear problems. EKFs are KFs that employ linear approximations for either the nonlinear target dynamics (i.e., nonlinear system models) or nonlinear measurement equations (i.e., measurements that are nonlinear functions of the state vector) or both. These filters are based on Taylor series expansions of the exact nonlinear equations, where only the linear terms of the expansion are preserved. The equations can be linearized about the state vector or the measurement vector, depending on the application. The EKF equations are very similar to those of the KF, except that the linearized versions of nonlinear expressions are used.

Due to the nonlinear aspect of the problem being solved, simple discrete-time or sampled-data forms cannot always be used except for high filter update rates (where nonlinearities may be less pronounced). Instead, numerical integration is often required to accurately accommodate these effects (e.g., Runge Kutta-type numerical integration methods). Likewise, Jacobians are used for transforming between coordinate systems that are related by nonlinear expressions. Equations (7.10) and (7.12) are the expressions most affected by this requirement due to nonlinear dynamics and measurement-to-state mappings.

Interacting Multiple-Model (IMM) Filters. Interacting multiple-model (IMM) filters are based on the previously described KFs and EKFs, except that instead of a single model (embodied, e.g., by the state transition matrix, Φ, multiple models are employed, with the output being a blend of the solutions of the M individual filters.

Heuristically, a bank of KFs tuned to different models would seem to be an optimal solution as long as at least one of the models closely approaches the underlying process. The problem becomes on of selecting the "right" filter outputs. IMM filters employ blending probabilities to combine the M filter outputs for the optimal solution.

The equations that describe the IMM state and covariance estimates are:

$$\hat{x}\left(k|k\right) = \sum_{j=1}^{r}\hat{x}^{j}\left(k|k\right)\mu_{j}\left(k\right),$$
(7.14)

where $\hat{x}^{j}(k|k)$ is the updated state of the jth filter and is the probability that the underlying target model is that of the jth case, and:

$$P\left(k|k\right) = \sum_{j=1}^{r}\mu_{j}\left(k\right)\left\{ P^{j}\left(k|k\right) + \left[\hat{x}^{j}\left(k|k\right) - \hat{x}\left(k|k\right)\right]\left[\hat{x}^{j}\left(k|k\right) - \hat{x}\left(k|k\right)\right]^{T}\right\},$$
(7.15)

where $P^{j}(k|k)$ is the covariance matrix associated with the jth filter.

IMM filters result in higher computational requirements compared with the KFs or EKFs by a factor of approximately M when M models are employed. Since these filters only propagate each filter one step into the future (i.e., a single hypothesis per model), computer throughput and memory is constant, unlike the MHT approach where throughput and memory usage increases exponentially with each filter update, that is, growing as approximately kN^L where N is the number of hypotheses and L is the number of filter updates. As can be imagined, even when N and L are limited via the pruning of unlikely hypotheses, the IMM approach is much less computationally intensive. Alternately, while the MHT approach is optimal (and perhaps near optimal when reasonable pruning is performed), the IMM is suboptimal. However, when all things are considered, IMM filters are attractive solutions when a single model is inadequate for the types of targets to be tracked, or for different phases of a targets trajectory, such as for ballistic missiles.

IMM-JPDA Filters.   In terms of comparing the MHT and IMM approaches, the aggregate IMM-JPDA is more of an apples-and-apples comparison. In this tracking alternative, the JPDA is used for the data association and the IMM is used for state estimation. Due to its relative efficiency compared with MHT methods, the IMM-JPDA filtering approach is a nice alternative, both in terms of implementation complexity and computer throughput usage. Alternatively, when needed in stressing data association environments where use of multiple models is beneficial, MHT can also be used with IMM.

7.2.3   Classification, Discrimination, and Identification

The following sections describe algorithms used to perform the constituent elements of CDI processing.

7.2.3.1 Target Feature Extraction.
These types of algorithms obtain desired target features from raw radar measurements and tracking state vector estimates. There are two broad classes of target features: kinematics and signature features.

Kinematics Features.   These features are extracted from or computed from track state vectors. Common kinematics features include target speed and acceleration, altitude and altitude rate and acceleration. Two sub-functions of target extraction are feature computation and feature conditioning.

Feature computation for kinematics consists of computing quantities not explicitly contained in the target state vector. Common examples include altitude-based features. Feature conditioning typically entails some type of filtering or smoothing to reduce feature noise. Since the state vectors quantities are smoothed by definition, conditioning is usually applied only to features computed from kinematics, especially when particularly noisy due to nonlinear operations (e.g., trigonometric functions, inverses, exponential functions).

Signature Features.   Signature features include those based on target amplitude such as radar cross section (RCS) and related quantities. Since all signature features are primarily computed quantities, feature conditioning is very important. Signature features can be powerful discriminants when kinematics are not distinct among target classes, that is, overlap from class to class.

Feature Means and Covariances.   Another key element of feature extraction is computing estimates of feature means, error variances, and correlation coefficients, that is, the elements of an error covariance matrix. Ideally, these covariance matrices are functions of measurement-induced errors and target-dependent errors associated with different target classes. Feature conditional means are the expected value of a given feature, conditioned upon a specific target type:

$$\mu_{ij} = E\{f_i|c_j\},\qquad(7.16)$$

where f_i and c_j are the ith and j^{th} feature and target class, respectively. The feature error covariance matrix is defined as:

$$M = E\{\tilde{f}\tilde{f}^T\} = \begin{bmatrix} \sigma_{11}^2 & \rho_{12}\,\sigma_1\,\sigma_2 & \cdots \\ \rho_{12}\,\sigma_1\sigma_2 & \sigma_{22}^2 & \cdots \\ \rho_{1N}\,\sigma_1\,\sigma_N & \cdots & \sigma_{NN}^2 \end{bmatrix}.\qquad(7.17)$$

When statistical classifiers such as Bayes' algorithms are used, some form of equation (7.16) and (7.17) are required as explicit elements of the classifier, and in the Bayes' case they are necessary database quantities. When decision trees are employed for classifiers, these quantities can be used to derive decision boundaries.

7.2.3.2 Target Classifiers. The following subsections describe some common classifier algorithms.

Bayes' Classifier.   This ubiquitous statistical classifier is based upon Bayes' rule of conditional probability [3, 4]:

$$P\left(c_j \middle| f_i\right) \; = \; \frac{P\left(f_i \middle| c_j\right) P\left(c_j\right)}{\displaystyle\sum_{k=1}^{M} P\left(f_i \middle| c_k\right) P\left(c_k\right)}, \tag{7.18}$$

where $P(c_j|f_i)$ is the probability of target class j given that feature i is measured, $P(f_i|c_j)$ is the conditional probability of feature i occurring given that c_j is the underlying target class, and $P(c_j)$ is the class prior probability (i.e., the probability of class j occurring out of all J classes).

The two conditional probabilities in equation (7.19) are also referred to as a posteriori (or posterior) and feature probabilities. The J posterior probabilities are the classifier's outputs, and the feature means and class probabilities are elements of the classifier database. The feature probabilities are computed based on the underlying probability density, the feature means, and the error covariance matrix, defined in equations (7.16) and (7.17). For Gaussian-distributed feature probabilities, the form of the probability density is:

$$P\left(\vec{f} \middle| c_j\right) \; = \; \frac{1}{\left(\sqrt{2\pi}\right)^N |M|^{\frac{1}{2}}} \; e^{-\frac{\left(\vec{f}-\vec{\mu}\right)^T M^{-1}\left(\vec{f}-\vec{\mu}\right)}{2}}. \tag{7.19}$$

In equation (7.19), $\vec{f}$ is the measured feature vector, and $\vec{\mu}$ is the feature mean vector, and M is the feature error covariance matrix. When all features are independent and uncorrelated, equation (7.19) can be simplified to:

$$P\left(f_i \middle| c_j\right) \; = \; \frac{1}{\sqrt{2\pi}\,\sigma_{ij}} \; e^{-\frac{\left(f_i-\mu_{ij}\right)^2}{2\sigma_{ij}^2}}. \tag{7.20}$$

In real-world systems, a battle manager, command and control, or combat system will establish a minimum threshold test for the posterior probabilities to declare a target class. Equation (7.18) can be implemented recursively, where posterior probabilities can be used as prior probabilities on successive iterations. When posterior probabilities do not clearly indicate a single-class decision, the battle manager can defer its decision.

One requirement of a Bayes' classifier is that all possible target classes must be identified in the classifier database. This is critical since the Bayes' classifier will always compute posterior probabilities even if the correct class is not one of the target hypotheses (and at least one posterior probability will be the largest). Hence, an incomplete classifier database can lead to spurious and erroneous results if only the largest posterior probability is used as the metric to declare target classes. One solution for this problematic characteristic inherent in the Bayes' classifier is to define an unknown or "strange" class to accommodate nonidentified target classes. When this method is used, the strange class posterior probability can be used to assess the reasonableness of the apparent target class indicated by other posterior probabilities. Such an approach is a very important in effectively using Bayes' classifiers.

Given this limitation, the Bayes' classifier using equation (7.19) is the optimal linear classifier when assumed feature probability distributions match the true feature statistics. When other underlying probability densities are known a priori or can be estimated from measurements, these can be optimally used by the Bayes' classifier.

Dempster-Shafer Classifier. The Dempster-Shafer (D-S) classifier is a statistical classifier that uses the concepts of "evidence," "plausibility," and probability masses upon which to base target class decisions.

Analogous to the conditional probability used by the Bayes' classifier, the conditional probability mass of class A given features $v1$ and $v2$ can be expressed as:

$$m(A \,|\, v1,v2) = [m(A \,|\, v1)m(A \,|\, v2) + m(AvB \,|\, v1)m(A \,|\, 2) + m(A \,|\, v1)m(AvB \,|\, v2)]/D$$

$$(7.21)$$

The probability mass for $m(B \,|\, v1,v2)$ can be expressed in a similar fashion as in equation (7.21). Now consider the probability mass associated with classes A or B conditioned on the features:

$$m(AvB \,|\, v1,v2) = [m(AvB \,|\, v1) \, m(AvB \,|\, v2)]/D \qquad (7.22)$$

where D equals the sum of the numerators, and AvB means class A or B.

After all evidence has been considered, the D-S classifier needs a decision rule such as the plausibility of A given by:

$$P(A) = [m(A) + m(AvB)] / [m(A) + m(AvB) + m(B) + m(AvB)]. \qquad (7.23)$$

As described in [5, 6], the evidence leading to a decision is the probability masses associated with the candidate target hypotheses. Using the mass combination rules, such as that represented by equation (7.23), the plausibility of the underlying target classes can be computed.

Key differences between D-S and Bayes' are the use of unnormalized probabilities (i.e., the probability "masses"), and a probability distribution-free approach compared with the Bayes' classifier, which often assumes an underlying Gaussian probability distribution. Another difference is the ability to handle correlated features. Bayes' theory incorporates feature correlation information via the feature error covariance matrix, and specifically by the off-diagonal terms. D-S theory does not account for feature interdependencies. For radar applications, this can be a deficiency of the D-S classifier as compared with the Bayes' methods. Although the D-S classifier can be modified to account for correlated features, these adjustments are ad hoc in nature and are sub-optimal solutions compared with the Bayes' classifier.

Decision Tree Classifiers. One of the simplest classifiers is a decision tree with a fixed structure and decision rules. Decision trees are desirable when minimizing computer throughput is a strong consideration in classifier selection and feature statistics are not available or cannot be quantified. Decision trees can employ nonquantitative features and concepts such as "slow targets" versus "fast targets," or "short targets" versus "long targets," "manned targets" versus "unmanned targets," and similar "fuzzy" target-related attributes.

A key rule in designing decision trees is to employ the highest-quality features or those with the greatest discriminating capabilities early in the decision process and lower-quality or less discriminating features later in the tree. Figure 7.8 depicts a typical decision tree for use of target total energy (i.e., potential plus kinetic energies) to separate tactical ballistic missiles (TBMs) and air-breathing targets (ABTs).

Rule-Based Classifiers. Another type of target classifier is the rule-based algorithm class. These may or may not use quantitative features and can make

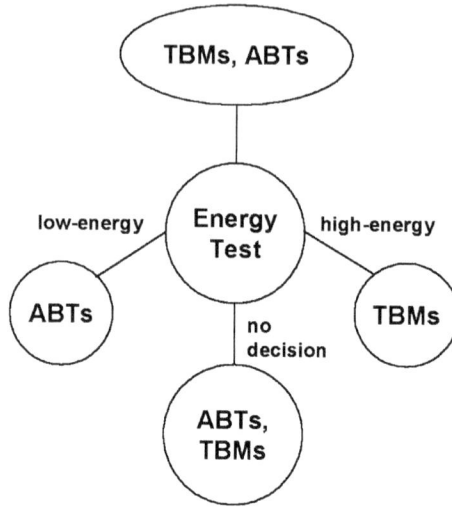

Figure 7.8 A Typical Decision Tree Classifier

"hard" or "soft" decisions, unlike simple decision trees that only make "hard" decisions (e.g., a target is in class A or B not, perhaps, in both classes). These rules are usually logical functions such as "if-then-else." An example of rule-based classifier constructs are:

$$\text{If \{speed is slow\}}$$
$$\text{Then \{target is a helicopter or UAV\}} \tag{7.24}$$
$$\text{Else \{target is a tank or ship\}}$$

or,

$$\text{If \{speed} > v_p\}$$
$$\text{Then \{target is jet-powered\}} \tag{7.25}$$
$$\text{Else \{target is propeller powered\}.}$$

As can be seen, either qualitative or quantitative rules can be employed. This provides the ability to use so-called fuzzy logic or neural-like processes.

Compound Classifiers. As one might expect, the classifiers discussed in the preceding sections are a sample of some commonly used ones; there are many more defined in [3, 4]. However, another possible classifier type is based on the combination of one or more of these (and others) algorithms to create a "compound" classifier.

Often, target classification processing might use a decision tree at a high level as the overall classifier structure. Then each node might employ different types of classifiers such as Bayes' or D-S or rule-based approaches. Since each of the classifiers discussed has strengths and weaknesses, the best solution is to use the right classifier for separating or classifying targets at the particular stage of classification processing based on performance, efficiency, and so on. This approach can yield a very powerful solution technique for target classification, discrimination, and identification problems encountered in radar applications.

7.2.4 Radar Hardware Control

The following subsections describe some commonly used radar hardware control algorithms and processing.

7.2.4.1 Waveform Control. These algorithms control the waveform generation and transmission functions. Both are typically associated with the control of a radar exciter or transmitter subsystem. Key data provided to exciters are waveform frequency, pulse length, modulation type and parameters, transmit time, initial phase, and so on. An example is specifying an uncoded (CW) 5 μs pulse, at 5 GHz, with a start time of 1 millisecond from the transmit execution time. Another example is a 1-millisecond up-chirp (LFM) with 1 GHz bandwidth at an operating frequency of 10.2 GHz, 1.5 milliseconds start time, and 0 degree initial phase offset. These parameters must be supplied to the exciter for each transmit action. Note that waveforms can also be coherent and non-coherent pulse trains necessary for pulse-Doppler operation, or to enable coherent integration or non-coherent integration, respectively.

7.2.4.2 Antenna Steering Control. These algorithms control the antenna steering, which can be mechanical, phase-steered, time-delay steering, or some combination of these methods. For fixed electronically steered arrays (ESAs), most steering commands are converted from azimuth and elevation into direction cosines u and v. Next, depending on the type of electronic steering employed, phase-shift, time-delay, or combinations of both can be commanded of the beam steering generator (BSG) hardware subsystem. These commands are required for each transmit and receive action.

For mechanically steered antennas, servo commands might consist of antenna slew rates, desired azimuth and elevation positions, and so on. Rotating antennas might require rotation rate commands. In use of these types of antennas, the dynamical models associated with an antenna pedestal or mount motion (e.g., defined by servo control system parameters, motors, and antenna mass) must be used to derive antenna pedestal/mount commands. These commands must be supplied at rates required by the antenna servo control system.

7.2.4.3 Receive Control. These algorithms are complementary to those used for waveform control. The radar receiver (or receivers) must be commanded to perform RF down-conversion and sampling to return signals within the desired range window (i.e., listen interval). Parameters include waveform frequency, pulse length (or waveform code), bandwidth, modulation, range window start time (relative to transmit time), range window extent, and A/D sampling rates, and so on. Additional commands relate to receiver processing such as de-ramp (or de-chirp) mixing or down-conversion needed for stretch-processed waveforms. Again, these commands are required for each and every receive action.

7.2.5   Radar Measurement Processing

These algorithms are typically referred to as returns handling and are described in the following sections.

7.2.5.1 Amplitude and Phase Calibration. A major purpose for returns processing is application of any calibration for amplitude and phase alignment. Often the calibration data are generated from "pilot-pulse" calibration processing that collects data at the signal processing output for waveforms injected into the front-end of the radar for the relevant range of system parameters. Other system-level calibration data are generated from data collected from tracked calibration targets (e.g., metric calibration satellites). This is usually based on absolute amplitude and phase accuracy.

7.2.5.2 Range Calibration. These algorithms adjust reported target range using calibration data such as "range zero" derived from pilot-pulse processing and satellite tracking, and so on. The time alignment necessary for multiple-pulse integration, on the other hand, is based on tracking data, and is a predicted time shift commanded of the exciter at the time of waveform generation.

7.2.5.3 Monopulse Calibration. Monopulse calibration adjusts the monopulse ratios (both real and imaginary components) reported by the signal processor for frequency-dependent, antenna mode-dependent, and waveform pulse length and bandwidth-dependent variations based on measured data. The measured data are used to extract antenna characteristics such as monopulse slope nonlinearities. The corrections are applied to the raw monopulse measurements provided by the signal processor.

7.2.5.4 Channel-to-Channel Alignment. For multichannel radars (i.e., those that employ monopulse, sidelobe blankers, sidelobe cancellers, or adaptive processing), it is necessary to ensure relative alignment of all channels with respect to time, amplitude, and phase. Again, this is typically performed based on pilot-pulse calibration processing-based data. The required accuracy of these alignments is driven by system-level requirements, such as specified jamming or interference cancellation.

7.2.5.5 Radar Cross Section Calibration. This amplitude correction, unlike channel-to-channel calibration, is an absolute calibration based on tracking a known RCS target such as a metric calibration satellite. These calibration data are frequency, waveform, and automatic gain control (AGC) dependent. Each return's amplitude used for RCS or SNR computations is adjusted by the appropriate RCS calibration factor prior to its use.

7.2.6 Signal Processing

These algorithms include pulse matched filtering, detection processing, monopulse processing, interpolation and peak detection, and coherent and non-coherent multiple-pulse integration. These are described in more detail in the following subsections.

7.2.6.1 Matched Filtering. The matched filtering algorithms are the optimal waveform processing under the criterion of maximizing output SNR. Therefore, radar processing nearly always employs matched filters prior to detection processing. Since most radars use linear frequency modulation (LFM or chirp modulation) waveforms, the digital matched filter typically falls into one of two classes: all-range digital pulse compression and "stretch" processing (also known as spectrum analysis).

The former approach is used for low and medium bandwidth applications such as search and tracking. The usual form of digital pulse compression is depicted in the block diagram of Figure 7.9.

As can be seen from Figure 7.9, the basic signal processing "building block" is the fast Fourier transform diagram shown in Figure 7.10. Using this building block, any type of digital matched filter can be synthesized.

Stretch processing consists of de-ramp (or de-chirp) down-conversion in the receiver, followed by spectrum analysis performed by the signal processor. Hence, the signal processing portion of stretch processing can be synthesized by tailoring the building block in Figure 7.10 to yield the block diagram in Figure 7.11.

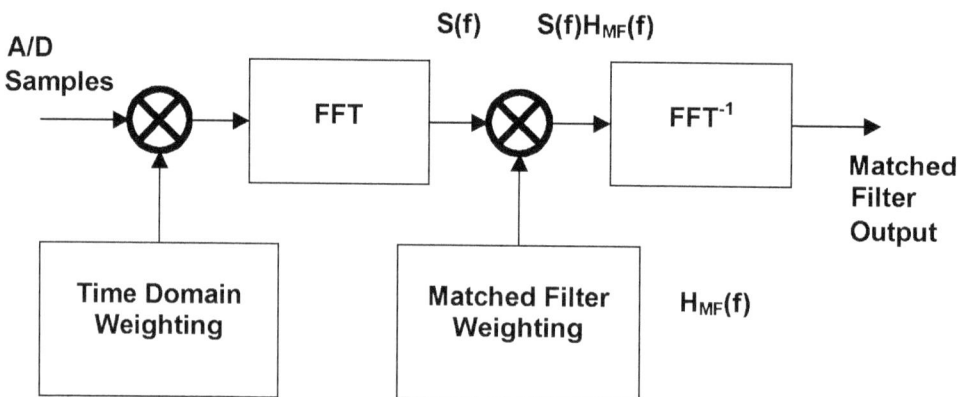

Figure 7.9   Digital Pulse Compression Processing

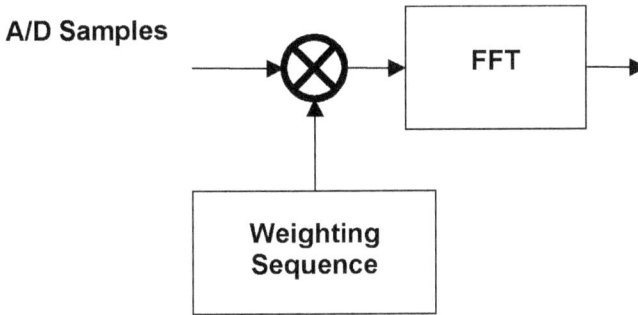

Figure 7.10   Basic Fast Fourier Transform "Building Block"

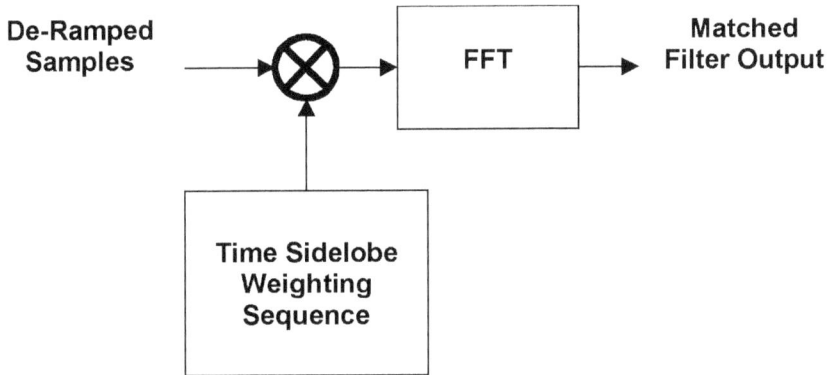

Figure 7.11   Spectrum Analysis Portion of Stretch Processing

7.2.6.2 Detection Processing. Detection algorithms implement the thresholding of match filtered output for the purpose of declaring target returns [7]. Typically, there are several detection thresholding approaches, including:

• Noise threshold

• CFAR threshold

• Linear detector

• Logarithmic detector

• RCS threshold.

These algorithms are described in the following subsections.

Noise Threshold.   This is the simplest threshold determination algorithm. A sequence of noise samples is collected as a function frequency, bandwidth, azimuth and elevation angles, etc., to estimate a background noise average using:

$$\eta_{AVE} \ = \ \frac{1}{N_S} \sum_{i=1}^{N_S} n_i \ . \tag{7.26}$$

By selecting a large enough number of samples to compute equation (7.26), an arbitrarily accurate noise estimate can be obtained. This noise estimate is used in the threshold test of the form:

$$s_{OUT}(t) \ \overset{H_1}{\underset{H_0}{\gtrless}} \ -\ln P_{FA} \ \eta_{AVE}, \tag{7.27}$$

where H_1 is the hypothesis that a target is present, H_0 is the null hypothesis, that is, only noise is present, and P_{FA} is the probability of false alarm.

CFAR Threshold.   The constant false alarm rate (CFAR) algorithm uses fewer noise samples, leading and lagging the signal under test, to compute a background noise estimate as defined by equation (7.26). Due to the smaller number of samples averaged, the estimation error is larger, which leads to a loss when using this threshold approach as compared with the noise threshold described under "Noise Threshold."

The threshold test is identical to equation (7.27), except that the right-hand side is replaced with one of the following:

$$
\begin{aligned}
& -\ln P_{FA} \ \eta_{AVE-LEAD} \\
& -\ln P_{FA} \ \eta_{AVE-LAG} \\
& -\ln P_{FA} \left(\frac{\eta_{AVE-LEAD} \ + \ \eta_{AVE-LAG}}{2} \right), \\
& -\ln P_{FA} \ \max\left(\eta_{AVE-LEAD}, \eta_{AVE-LAG} \right) \\
& -\ln P_{FA} \ \min\left(\eta_{AVE-LEAD}, \eta_{AVE-LAG} \right)
\end{aligned}
\tag{7.28}
$$

or any of the above, with the M largest samples (i.e., range-cells) censored or deleted from the noise average computation.

As stated above, there is a CFAR loss in detection, relative to the theoretical (ideal) noise threshold. It varies with numbers of samples used in the leading and lagging averages, as well as whether linear or logarithmic processing is employed, after reference [8]. Figure 7.12 depicts a basic CFAR processing block diagram.

Linear and Logarithmic Detectors. Linear and logarithmic detectors are two commonly used detection schemes (a third is square-law detectors). Linear detectors are those defined by equations (7.27) and (7.28). These are more prevalent than their logarithmic counterparts today, since logarithmic processing was used in situations where large dynamic range was required (e.g., in severe clutter environments) and dynamic range is much less of a problem in today's radar hardware.

Following a matched filter by a logarithmic converter compresses the dynamic range allowing a larger range of small to large signals to be processed. However, the downside of such processing is the potential for suppression of small targets

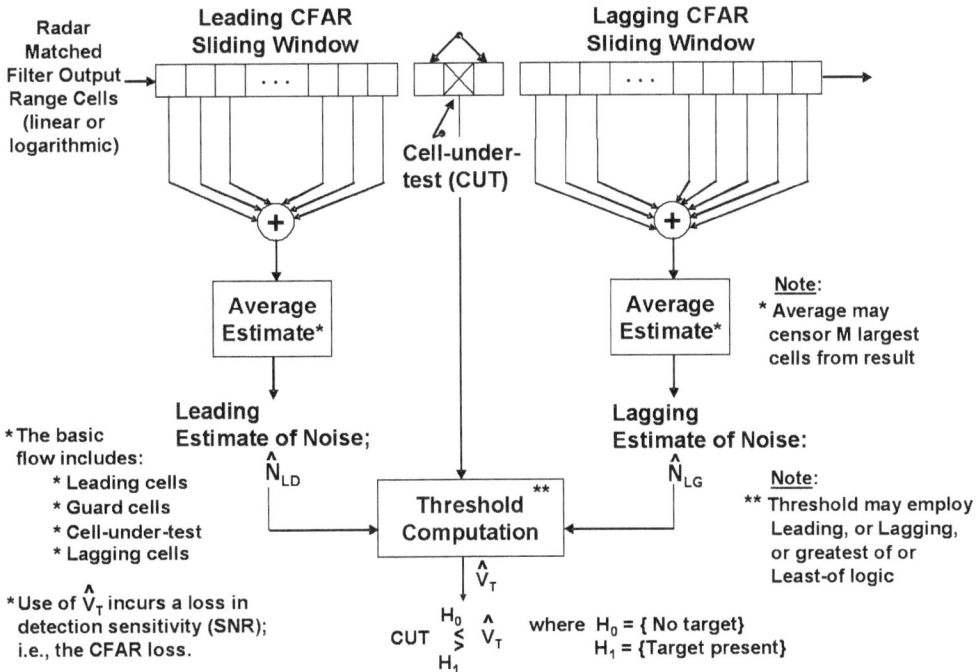

Figure 7.12 CFAR Processing Block Diagram

in the vicinity of large targets due to the inherent nonlinear response of a logarithm processor. Hence, there is a possible loss of several dB for small targets when near large targets for logarithmic detectors (e.g., a reentry vehicle [RV] or warhead near a rocket booster encountered in missile defense scenarios).

The analog to equation (7.27) for a logarithmic detector is:

$$
\begin{aligned}
\log_2\left[s_{OUT}(t)\right] &\overset{H_1}{>} \log_2\left[-\ln P_{FA}\right] + \log_2 \eta_{AVE} \\
\log_2\left[s_{OUT}(t)\right] &\overset{H_0}{\leq} \log_2\left[-\ln P_{FA}\right] + \log_2 \eta_{AVE}
\end{aligned}
\tag{7.29}
$$

Although the base 2 logarithm is used in equation (7.29), any arbitrary base can be employed. Use of the base 2 logarithm in equation (7.29) is convenient for digital processing (i.e., using binary arithmetic), while use of base 10 logarithms provides detection thresholds in units of decibels by simply multiplying by a factor of 10. As for the log-CFARs, logarithmic detectors yield a loss in detectability when compared to the linear test defined by equation (7.27).

RCS Threshold.   At times, it is desirable to filter out or screen small RCS targets (e.g., birds) as candidates for detection. An RCS threshold can be used for this purpose, where the predicted amplitude for a specified RCS can be computed as:

$$
\eta_{RCS} = \sqrt{\frac{k_{RADAR}\, RCS}{R^4}},
\tag{7.30}
$$

where k_{RADAR} is a radar gain constant (a function of the radar range equation parameters, including antenna transmit and receive gains, wavelength, and peak transmitter power), and R is the slant range where the threshold test is to be applied. η_{RCS} can be substituted for the term on the right in equation (7.27), that is, $-\ln P_{FA}\eta_{AVE}$ for this purpose.

7.2.6.3 Monopulse Processing. These signal processing algorithms are used to extract target angle estimates (i.e., azimuth and elevation angles, or direction cosines u and v in the case of fixed phased-array antennas) from Σ, α, β monopulse channel voltages. The basic first-order linear scaling is given by:

$$\varepsilon_{AZ} = \mathrm{Re}\left\{ \frac{\theta_3}{k_{m-AZ}} \frac{\alpha}{\Sigma} \right\}$$

$$\varepsilon_{AZ} = \mathrm{Re}\left\{ \frac{\theta_3}{k_{m-EL}} \frac{\beta}{\Sigma} \right\}, \qquad (7.31)$$

where θ_3 and k_m are the antenna 3 dB receive beamwidth and monopulse slope, and ε_{AZ} and ε_{EL} are the azimuth and elevation angle errors (i.e., relative to antenna electrical boresight), respectively, and $\mathrm{Re}\{\cdot\}$ is the real part of the complex-valued monopulse ratios. The imaginary counterparts to the quantities in equation (7.31) can be used to edit unresolved target data to prevent corruption of tracking or target classification or identification.

7.2.6.4 Coherent and Non-Coherent Integration. These two algorithms are used to improve target detectability by adding multiple target returns prior to the thresholding operation described in Section 7.2.6.2. Coherent integration uses amplitude and phase measurements to add multiple, pulse-to-pulse coherent target returns to yield increased return amplitude via:

$$s_{CI}(t) = \sum_{i=1}^{N} \left[s_I(i) + s_Q(i) \right]. \qquad (7.32)$$

Alternatively, when returns are not coherent from pulse to pulse, return magnitudes are added, rather than the vector addition performed in equation (7.32), to yield:

$$s_{NCI}(t) = \sqrt{\sum_{i=1}^{N} \left[s_I(i)^2 + s_Q(i)^2 \right]}. \qquad (7.33)$$

7.2.7   Calibration and Alignment

Two major types of calibration and alignment are built-in processing and tracked object-based methods. They are described in the following subsections.

7.2.7.1 Built-In Calibration and Alignment. Radar designs can incorporate several built-in calibration techniques. One common approach is often referred to as "pilot pulse" processing. This algorithm involves the use of RF signal injection into all receiver channels and computing the time, amplitude, and phase corrections necessary to align them. These data are stored, and subsequently are applied to the different channel output signals. The waveforms used for pilot pulse processing cover the full range of operating frequencies, pulse lengths, ACG settings, bandwidths, and so on, and the resulting outputs are processed to compile the desired calibration coefficients for each of the distinct combinations of parameters described above.

Additional types of built-in calibration and alignment include such techniques as obtaining samples of transmit and receive signal to use for adjusting certain radar measurements. Also, both antenna subarray and array tests can be used to verify antenna component alignments.

7.2.7.2 Tracked Object-Based Calibration and Alignment. These types of calibration are performed at the system level and involve tracking real-world targets such as balloons and calibration satellites. Tracking-based calibration and alignment covers the parts of the radar processing chain not covered by built-in methods such as pilot pulse processing. These methods require the tracking of balloons or satellites with known and calibrated RCS values, as well as tracking satellites with well-known trajectory data. The former type of tracked objects can be used to collect RCS measurements as a function of operating frequency, waveform bandwidth, and so on to establish RCS calibration coefficients, as well as be used to collect waveform response data to compute matched filter replicas. The latter type of target, namely, metric calibration satellites, can be used to measure range, and azimuth and elevation angles and compute calibration coefficients for these measurements as a function of operating frequency, waveform bandwidth, and so on. Radars that have very accurate metric accuracy requirements will use these techniques to correct measurements for repeatable hardware-induced variability and biases (i.e., systematic errors). These corrections are applied by the measurement processing algorithms described in Section 7.2.5.

7.2.8 Adaptive Processing

The following algorithms represent those used to mitigate the effects of interference and intentional barrage noise jamming. There are three major classes of these techniques [8, 10]:

• Sidelobe cancellers (single and multiple)

• Adaptive arrays

• Space-time adaptive processing.

The theory describing adaptive arrays can also be applied to adaptive Doppler-based signal processing. These three techniques are described in the following subsections.

7.2.8.1 Sidelobe Cancellers (SLCs). Single sidelobe cancellers (SLCs) and multiple SLCs (or MSLCs) are the most commonly used interference and jamming cancellations techniques. The primary difference between the two approaches is that the MSLC processes M auxiliary antenna channels concurrently and can cancel up to M interference or jamming sources.

A basic block diagram of an MSLC employing M auxiliary antennas (i.e., M canceller "loops") is depicted in Figure 7.13.

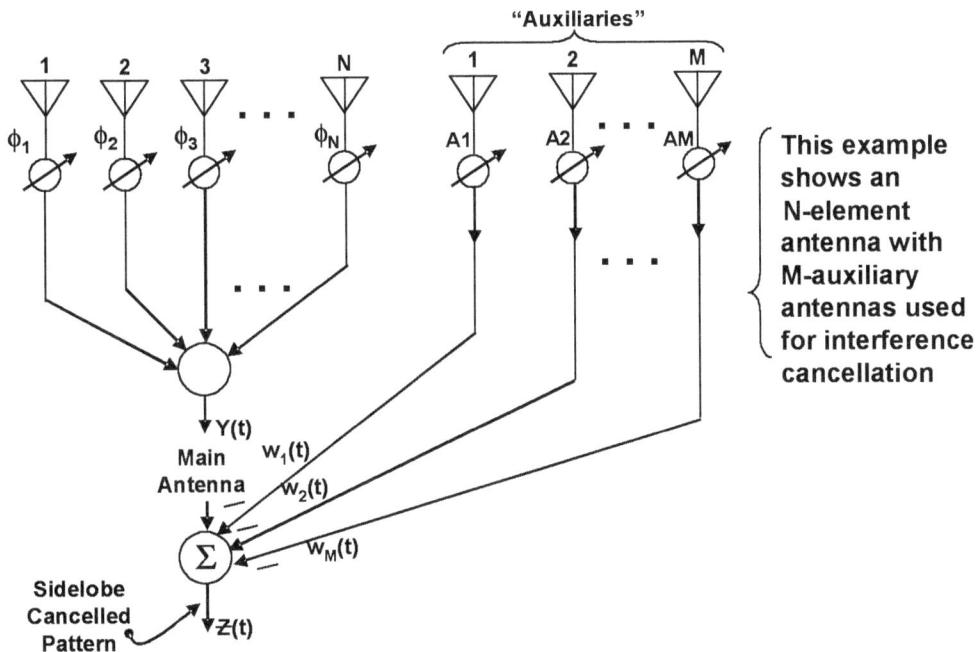

Figure 7.13   Block Diagram of MSLC with M Degrees-of-Freedom

The SLC concept is to estimate the jamming or interference using an auxiliary antenna and to subtract it from the main antenna-pattern response. The output of the SLC-based system is then:

$$z(t) = j(t)\left[g_m(\theta_J) - w^* g_a(\theta_J) \exp\left(-j2\pi f_0 \tau_j\right)\right] ,$$ (7.34)

where:

$\quad j(t)$ = complex envelope of jammer voltage

$\quad g_m(\theta_J)$ = main antenna gain in jammer direction (θ_J)

$\quad w^*$ = conjugate of complex weight, w

$\quad g_a(\theta_J)$ = auxiliary antenna gain in jammer direction (θ_J)

$\quad f_0$ = operating RF frequency

$\quad \tau_j$ = propagation delay between main and auxiliary antenna.

For a single SLC the weight is given by:

$$w = \frac{E\left\{y(t)j(t)^*\right\}}{E\left\{j(t)j(t)^*\right\}} = \rho \frac{P_{MAIN}}{P_{AUX}} ,$$ (7.35)

where ρ is the correlation coefficient between the main and auxiliary channels, and P_{MAIN} and P_{AUX} are powers in the main and auxiliary channels, respectively.

The effect of the sidelobe canceller is to minimize the resultant antenna pattern gain in the direction of the jamming source. MSLCs operate in a similar manner, except the M weights are solved using a vector-matrix form of equation (7.35):

$$\vec{w} = E\left\{\vec{\varepsilon}\vec{\varepsilon}^{*T}\right\}^{-1} E\left\{y\vec{\varepsilon}^*\right\} ,$$ (7.36)

where the first term on the right is the error covariance matrix for the M auxiliary channels, and the second term is the cross-correlation vector of the main channel and the M auxiliary channels. If there are fewer than M jamming sources, then more than 1 degree-of-freedom will be used to cancel a single jammer.

7.2.8.2 Adaptive Arrays. Adaptive arrays are the logical extension of MSLCs to utilize larger degrees-of-freedom to mitigate interference and jamming. Here, instead of auxiliary antennas, the adaptive array employs N elements or subarrays of the main antenna array. A block diagram of an adaptive array appears in Figure 7.14.

Although the adaptive array and MSLC block diagrams differ, the processing is quite similar. The adaptive weight vector is given by:

$$w_{OPT} = R^{-1} s ,$$ (7.37)

where R is the error covariance matrix, and s is the cross-correlation vector defined as:

$$s = E\left\{ y y_n^* \right\} .$$ (7.38)

Alternatively, s can be considered the desired steering vector. As can be seen, equations (7.36) and (7.37) are quite similar in form.

The key to either the MSLC or adaptive array approaches is computation of the error covariance matrix and cross-correlation vector. There are several methods available to obtain these quantities. The first is to estimate the covariance matrix by averaging a number of measurements, that is, computing:

$$R = \frac{1}{N_S} \sum_{i=1}^{N_S} y\, y^{*T} .$$ (7.39)

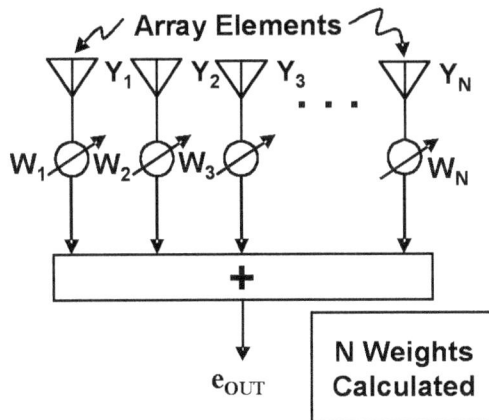

Figure 7.14 Adaptive Array Block Diagram

This technique is referred to as the sampled matrix inversion (SMI) approach. As can be expected, since w is a random process, the averaging process in equation (7.39) reduces the estimation error. Therefore, as N_s approaches ∞, w approaches w_{OPT}. A rule of thumb is that $N_s = 2N$ reduces the error relative to w_{OPT} to about 3 dB, and $N_s = 4N$ reduces the error to about 1 dB, reference [9].

Other approaches to solving for the weight vector involve factoring the covariance matrix into upper and lower-triangular forms (upper-diagonal [U-D] factorization). This allows solving for without explicitly computing the matrix inverse as defined in equation (7.37) for the SMI method. Another approach is the recursive least-squares (RLS) technique. RLS computes a least-square error solution for the weight vector using an iterative technique. Random search techniques can also be employed to select a weight vector by evaluating residual jammer levels for values of w selected using a random algorithm.

Two commonly used methods for weight vector computation are the SMI and U-D factorization (and similar matrix approaches to solving N equations and N unknowns) types. For the same set of conditions, both methods yield similar performance, although the latter approach is more numerically stable or well-conditioned (e.g., when the ratio of small-to-large jamming or interference sources is large).

7.2.8.3 Space-Time Adaptive Processing. STAP is predominantly employed by airborne radars and space-based radars, but the concepts can apply equally well to all radars that require adaptation in both spatial and temporal dimensions due to severe environmental conditions. The spatial adaptation is identical to that for the adaptive arrays in Section 7.2.8.2. However, the temporal (or frequency domain) adaptation is conceptually different. In many applications, clutter needs to be filtered spectrally when spatial filtering alone does not suffice.

The time or frequency methods are identical to an adaptive moving target indicator (AMTI) or adaptive Doppler processing. A STAP processor working in the time domain is depicted in Figure 7.15, courtesy of E. Parsons.

As can be seen, STAP has N spatial weights and K temporal weights providing a total of NK degrees-of-freedom (DoF). The additional DoF can be used to cancel wideband interference sources, as wideband interference requires more than

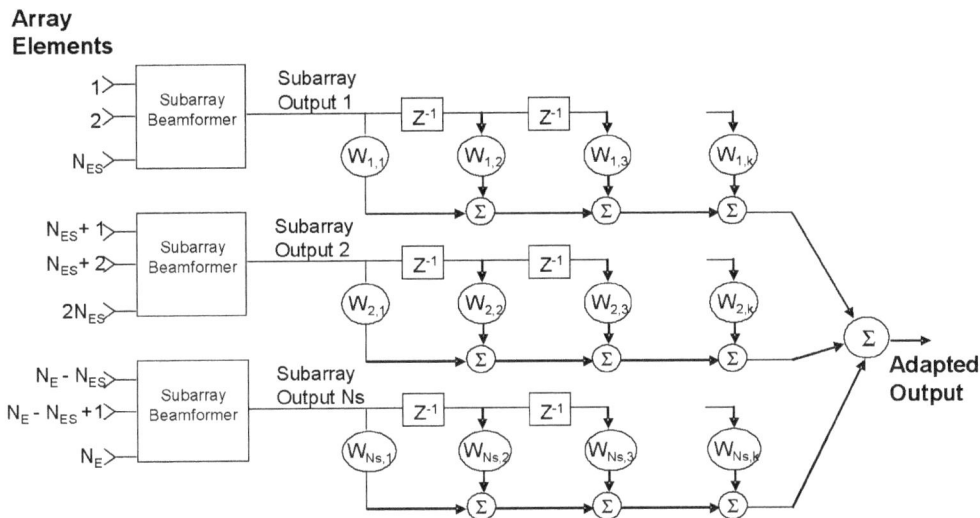

Figure 7.15 STAP Block Diagram

one DoF per source. The STAP approach can also be used to cancel clutter by synthesizing clutter-mitigating filters (or pre-whitening filters) for target detection purposes.

Computation of the NK weights can be performed analogously to the adaptive array, except that higher-order covariance matrices must be estimated instead of the $N \times N$ covariance matrix as for the latter case. Similar covariance estimation algorithms to those described in Section 7.2.8.2, including the SMI, U-D factorization, RLS, and random methods can be used for STAP weight computation.

7.2.8.4 Digital Beam Forming (DBF). Although not an adaptive processing algorithm, DBF does offer features that enable these algorithms. The basic DBF radar architecture is depicted in Figure 7.16. Fundamentally, a receiver is placed behind each antenna element, super-element (i.e., groups of elements driven by a common source), or subarray, depending the DoF desired, with outputs converted to digital format (this receiver is referred to here as a "digital" receiver). DBF replaces the usual "hardwired" sum and difference beam formers (implemented via hardware, e.g., waveguides, coaxial cables, RF circuit boards) with as many distinct antenna combiners or "channels" as desired. These digitally formed beams or processing channels can be used in the usual

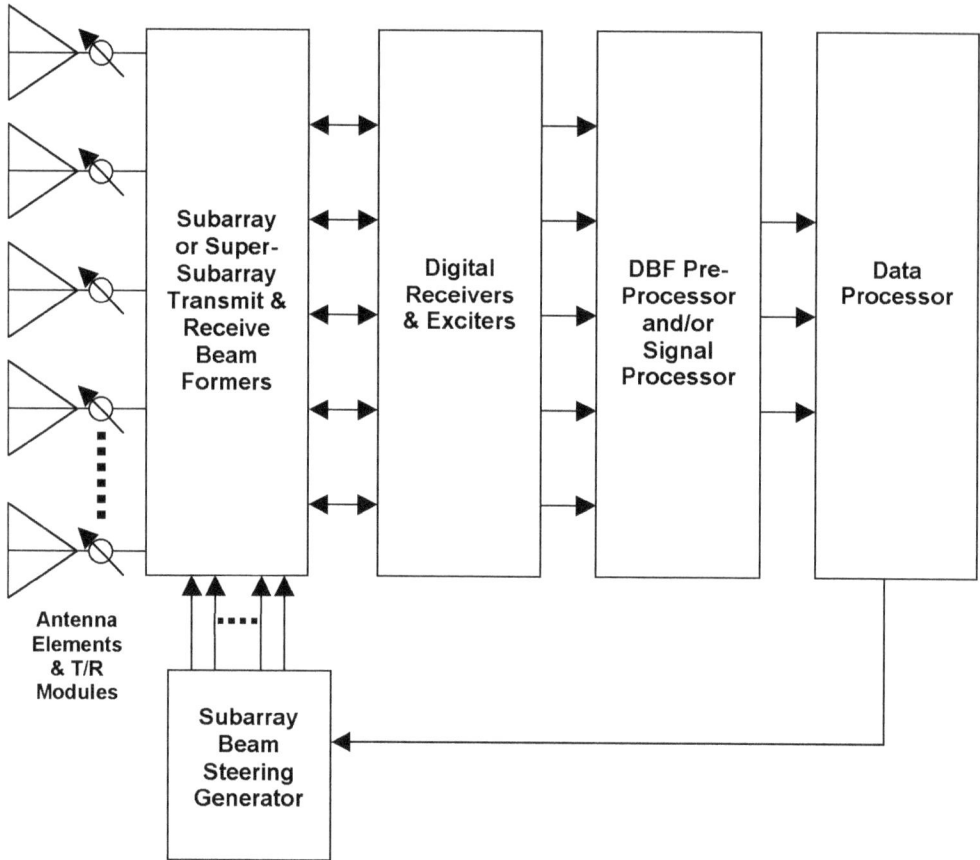

Figure 7.16   DBF Radar Architecture

manner to form monopulse antenna patterns (Σ, α, β) needed for tracking, or to create partitioned antenna apertures or multiple simultaneous receive beams. The system-level cost of using DBF is the need for separate receiver and signal processing chains for each DoF in the multiple-channel architecture.

7.2.9   Statistical Detection and Estimation

Detection algorithms are described in Section 7.2.6.2. Although these, too, are statistical in nature, they are limited to simple hypotheses concerning the target and its environment. Statistical detection and estimation as described in the following subsections represents the more general detection and estimation algorithms that relax the underlying assumptions in Section 7.2.6.2 to yield more general algorithms and processing [7, 11].

7.2.9.1 Generalized Likelihood Ratio Test.

The generalized likelihood ratio test (GLRT) is a useful technique for target detection purposes when there are unknown aspects of the problem, including target and environmental parameters [7, 11]. When any of the necessary parameters of the likelihood ratio test (LRT) are unknown random variables, or can be modeled as such, the GLRT allows for use of estimates of the unknown quantities. This is analogous to the processing performed by the MSLC and adaptive array algorithms in Section 7.2.8, which estimate unknown error covariance matrices to compute the adaptive weights.

The likelihood ratio (LR) can be stated as:

$$\lambda(y) = \frac{p(y \mid H_1)}{p(y \mid H_0)}. \tag{7.40}$$

The two target hypotheses in equation (7.40), H_0 and H_1, represent the null (i.e., no target or noise-only) and target-plus-noise hypotheses, respectively. If the probabilities specified in the LR are a function of unknown parameters (e.g., target amplitude, RCS fluctuation characteristics, frequency, phase), then they must be estimated prior to performing equation (7.40). When this is the case, the generalized likelihood ratio (GLR) is used instead of the LR.

Consider the case where the conditional probabilities given H_0 and H_1 are defined as Gaussian-distributed random variables:

$$p(y \mid H_0) = \frac{1}{\sqrt{2\pi\sigma_n^2}} \exp\left\{ -\frac{y^2}{2\sigma_n^2} \right\}$$

$$p(y \mid H_1) = \frac{1}{\sqrt{2\pi\sigma_s^2}} \exp\left\{ -\frac{(y-A)^2}{2\sigma_s^2} \right\} + \frac{1}{\sqrt{2\pi\sigma_n^2}} \exp\left\{ -\frac{y^2}{2\sigma_n^2} \right\} \tag{7.41}$$

Substituting equation (7.41) into (7.40) yields the LR:

$$\lambda(y) = \frac{p(y \mid H_1)}{p(y \mid H_0)} = \frac{\frac{1}{\sqrt{2\pi}\sigma_s}\exp\left\{-\frac{(y-A)^2}{2\sigma_s^2}\right\} + \frac{1}{\sqrt{2\pi}\sigma_n}\exp\left\{-\left[\frac{y^2}{2\sigma_n^2}\right]\right\}}{\frac{1}{\sqrt{2\pi}\sigma_n}\exp\left\{-\left[\frac{y^2}{2\sigma_n^2}\right]\right\}}.$$

$$(7.42)$$

Now, consider the case where signal amplitude, A, and the root mean square (RMS) signal and noise powers, σ_s^2 and σ_n^2, respectively, are unknown parameters and are modeled as random variables. Then the LR of equation (7.42) becomes the GLR given by:

$$\lambda(y) = \frac{p(y \mid H_1)}{p(y \mid H_0)} = \frac{\frac{1}{\sqrt{2\pi}\sigma_s}\exp\left\{-\frac{(y-\hat{A})^2}{2\sigma_s^2}\right\} + \frac{1}{\sqrt{2\pi}\sigma_n}\exp\left\{-\left[\frac{y^2}{2\hat{\sigma}_n^2}\right]\right\}}{\frac{1}{\sqrt{2\pi}\sigma_n}\exp\left\{-\left[\frac{y^2}{2\hat{\sigma}_n^2}\right]\right\}},$$

$$(7.43)$$

where, $\hat{A}$, $\hat{\sigma}_s^2$, and $\hat{\sigma}_n^2$ are the estimated values of A, σ_s^2, and σ_n^2, respectively. These estimates can be expressed as:

$$\hat{A} = E\{A\}$$

$$\hat{\sigma}_s^2 = E\{\sigma_s^2\}. \qquad (7.44)$$

$$\hat{\sigma}_n^2 = E\{\sigma_n^2\}$$

These estimates can be computed using the sample-average approaches described in Section 7.2.8 for adaptive processing algorithms. The resulting GLRT is depicted in Figure 7.17.

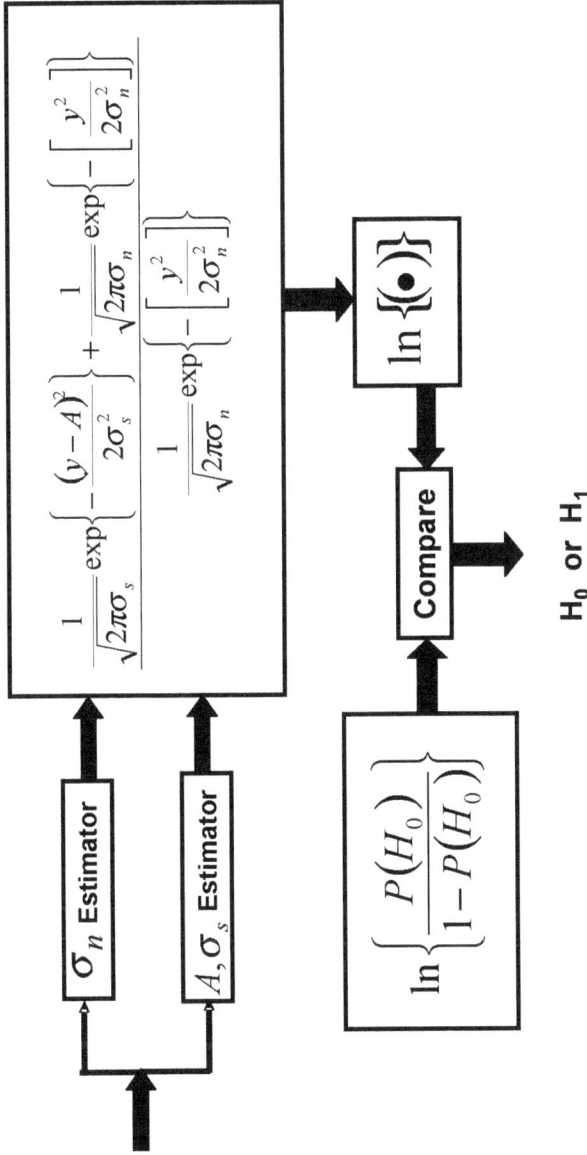

$$\frac{\dfrac{1}{\sqrt{2\pi}\sigma_s}\exp\left\{-\dfrac{(y-A)^2}{2\sigma_s^2}\right\} + \dfrac{1}{\sqrt{2\pi}\sigma_n}\exp\left\{-\left[\dfrac{y^2}{2\sigma_n^2}\right]\right\}}{\dfrac{1}{\sqrt{2\pi}\sigma_n}\exp\left\{-\left[\dfrac{y^2}{2\sigma_n^2}\right]\right\}}$$

σ_n **Estimator**

A, σ_s **Estimator**

$\ln\{(\bullet)\}$

Compare

$\ln\left\{\dfrac{P(H_0)}{1-P(H_0)}\right\}$

H$_0$ or H$_1$

Figure 7.17 GLRT for Gaussian Signal in Gaussian Noise Example

7.2.9.2 Statistical Estimation. Section 7.2.9.1 addresses one type of statistical detection algorithm that employs the estimation of target and environmental parameters as elements of the detection method. Use of other forms of statistical estimation is widespread in radar processing. Two previously discussed applications are statistical tracking filters (e.g., the Kalman filter) and statistical target classifiers (e.g., the Bayes' classifier). Statistical estimation techniques are described in more detail in [1–4, 7–11].

Innovations-Based Detection Algorithm (IBDA).   Another approach to statistical detection is the innovations-based detection algorithm (IBDA), after reference [11]. This general approach to statistical detection is predicated upon the predictor-corrector structure used for example in tracking algorithms, coupled with the GLRT described in Section 7.2.9.1.

   If the complex-valued baseband data are defined as the vector x, in the simplest case of a random target signal vector, s, in the presence of a general interference vector, c, two hypotheses can be postulated as:

$$
\begin{aligned}
H_0: & \quad x = c \\
H_1: & \quad x = s + c
\end{aligned}
\tag{7.45}
$$

where c and s are independent, zero-mean Gaussian random vectors with covariance matrices R_s and R_c. Hence, the conditional probabilities of the baseband data arising from the two hypotheses in equation (7.45) are:

$$
\begin{aligned}
P_0\left(x \mid H_0\right) &= \frac{1}{\pi^N \det\left(R_c\right)} \exp\left(-x^H R_s^{-1} x\right) \\
P_0\left(x \mid H_1\right) &= \frac{1}{\pi^N \det\left(R_s + R_c\right)} \exp\left(-x^H\left[R_s + R_c\right]^{-1} x\right)
\end{aligned}
\tag{7.46}
$$

Using equation (7.46) the log-LR is given by:

$$
l = x^H\left(R_c^{-1} - \left[R_s + R_c\right]^{-1}\right) x = \hat{s} \, R_c^{-1} x ,
\tag{7.47}
$$

where $\hat{s}$ is a smoothed MMSE of s defined as:

$$
\hat{s} = R_s\left(R_s + R_c\right)^{-1} x .
\tag{7.48}
$$

Equations (7.47) and (7.48) represent the standard predictor-corrector form alluded to above. As in other applications that require a matrix inverse, equation (7.48) can be implemented using upper-lower triangular factorization or other similar numerically stable solutions.

The test statistic of equation (7.47) is a general "innovation" expression of the optimal receiver structure. This is the source of the term IBDA. This formulation becomes simple if the data under both hypotheses can be expressed as autoregressive (AR) processes:

$$
\begin{aligned}
H_0: & \quad x(k) = \sum_{i=1}^{M_0} b_{M_0}(i)\, x(k-i) + e_0(k) \\
H_1: & \quad x(k) = \sum_{i=1}^{M} a_M(i)\, x(k-i) + e_1(k)
\end{aligned}
\tag{7.49}
$$

where the two error terms are independent, identically distributed, zero-mean Gaussian random variables with variances σ_0^2 and σ_1^2, respectively.

Now the log LR can be expressed as:

$$
L = \sum_{k=1}^{N} \left[\ln\left(\frac{d_0^2(k)}{d_1^2(k)} \right) + \frac{\left| e_0(k) \right|^2}{d_0^2(k)} - \frac{\left| e_1(k) \right|^2}{d_1^2(k)} \right].
\tag{7.50}
$$

By defining:

$$
\begin{aligned}
d_1^2(k) &= \sigma_1^2, \quad k=1,\dots,N-M \\
d_0^2(k) &= \sigma_0^2, \quad k=1,\dots,N-M_0 \\
\gamma_1^2(k) &= \begin{cases} 1, & 1 \le k \le N-M \\ d_1^2(k)/\sigma_1^2, & N-M < k \le N \end{cases} \\
\gamma_0^2(k) &= \begin{cases} 1, & 1 \le k \le N-M_0 \\ d_0^2(k)/\sigma_0^2, & N-M_0 < k \le N \end{cases}
\end{aligned}
\tag{7.51}
$$

equation (7.50) can be expressed as:

$$L = -\sum_{k=1}^{N}\left[\ln\left(\frac{\sigma_1^2}{\sigma_0^2}\right) + \ln\left(\frac{\gamma_1^2(k)}{\gamma_0^2(k)}\right) + \frac{|e_1(k)|^2}{\sigma_1^2\,\gamma_1^2(k)} - \frac{|e_0(k)|^2}{\sigma_0^2\,\gamma_0^2(k)} \right]. \quad (7.52)$$

In general, L is dominated by the terms with index less than $(N - \max\{M_0, M\})$ and the second term in equation (7.52) can be ignored. The yields for L:

$$L = -\sum_{k=1}^{N}\left[\ln\left(\frac{\sigma_1^2}{\sigma_0^2}\right) + \frac{|e_1(k)|^2}{\sigma_1^2} - \frac{|e_0(k)|^2}{\sigma_0^2} \right]. \quad (7.53)$$

Therefore, the general statistical detector using the IBDA and the AR models is the test that applies L from equation (7.53) to a threshold selected to ensure a specified probability of false alarm (P_{FA}).

7.3 REFERENCES

[1] Y. Bar-Shalom, *Multitarget–Multisensor Tracking: Principles and Techniques*, YBS, 1995
[2] S. Blackman & R. Popoli, *Design and Analysis of Modern Tracking Systems*, Artech House, 1999
[3] R. Duda, et al., *Pattern Classification*, 2nd Edition, Wiley-Interscience, 2000
[4] K. Fukunaga, *Introduction to Statistical Pattern Recognition*, 2nd Edition, Academic Press, 1990
[5] G. Shafer, *A Mathematical Theory of Evidence*, Princeton University Press, 1976
[6] P. Dempster, et al., *Classic Works on the Dempster-Shafer Theory of Belief Functions*, Springer, 2007
[7] H. Van Trees, *Detection, Estimation and Modulation Theory, Part 1*, Wiley-Interscience, 2001
[8] R. Nitzberg, *Radar Signal Processing and Adaptive Systems*, 2nd Edition, Artech House, 1999
[9] D. Manolakis, *Statistical and Adaptive Signal Processing*, Artech House, 2005
[10] R. A. Monzingo & T. M. Miller, *Introduction to Adaptive Arrays*, SciTech, 2003
[11] S. Haykin & A. Steinhardt, *Adaptive Radar Detection and Estimation*, Wiley, 1992

8

Interference Suppression Techniques

8.1 INTRODUCTION

Tactically deployed phased-array radars (PARs) must operate in environments where unintentional and intentional interference can potentially degrade performance. This chapter covers the concepts of interference suppression. References [1–3, 5, 6, 9] are excellent sources of background on this subject. The topics covered in this chapter include:

- Sources and types of electronic interference:
  - Unintentional
  - Electronic attack:
    - Sidelobe noise jamming
    - Mainlobe noise and repeater jamming
- Sidelobe blankers (SLBs)
- Sidelobe cancellers (SLCs)
- Multiple sidelobe cancellers (MSLCs)
- Adaptive processing

- Digital beam forming (DBF)

- Frequency agility and hopping

- Sector blanking (receive and transmit).

8.2   ELECTRONIC INTERFERENCE SOURCES

8.2.1   Unintentional Interference

Any in-band radio frequency (RF) source can be considered a potential interference source. Unintentional types include:

- Radio stations

- Television stations

- Cell-phones and repeaters

- Other radars

- Other in-band radiating devices.

These RF sources may be narrowband or of wider bandwidths relative to the radar operating bandwidths. However, the preponderance of real unintentional interferers fall into the former category.

   Due to the narrowband nature of most of these sources, they can often be mitigated with simple frequency censoring by the signal processor. This, of course, assumes that the interference does not saturate the radar front-end, that is, that the radar antenna and receiver have adequate linear dynamic range to accommodate the large RF signals levels.

8.2.2   Intentional Interference Sources

8.2.2.1 Sidelobe Noise Jammers. The most common form of electronic countermeasure (ECM) or electronic attack (EA) is broadband noise jamming directed in the direction of an antenna sidelobe to raise the front-end thermal noise floor and degrade the effective signal-to-noise ratio (SNR) or signal-to-interference ratio (SIR). The interferers can be at any range relative to the radar, air or space-borne, and can exhibit a wide range of power levels. Typically, jammers of this type spread their energy across the entire operating band of a radar.

8.2.2.2 Mainlobe and Repeater Jammers. A second major type of intentional interference is directed at the antenna's mainlobe. Mainlobe noise jamming can be similar in nature to sidelobe noise jamming discussed in Section 8.2.2.1 and may be located physically on the target or, more likely, on a platform accompanying the target. The former is referred to as a self-screening jammer, and the latter type is called an escort jammer. Again, a wide range of jammer power levels can be employed, and bandwidths will usually match or exceed the radar's operating bandwidth.

A repeater jammer's purpose is to "spoof" the radar, or to produce "decoy" radar echoes, that is, false targets intended to overload the detection and tracking capacity of a radar. Alternatively, a repeater's purpose might be either "range-gate" or "velocity-gate" stealing (i.e., to defeat the tracking function). Sophisticated repeaters can also simultaneously vary an echo's apparent range and apparent Doppler.

8.3 ELECTRONIC PROTECTION (EP) OR COUNTER-COUNTER MEASURES (ECCM)

The following sections describe a number of techniques available to the radar designer to mitigate the deleterious effects of interference.

8.3.1 Sidelobe Blanker

A sidelobe blanker is not really an electronic counter-counter measure (ECCM) but rather a method of editing antenna-sidelobe returns that can potentially confuse the tracking function, including the data association algorithms described in Chapter 5. The fundamental concept is to apply a test to all candidate target echoes to determine whether it originates in the antenna mainlobe or from a sidelobe. Figure 8.1 illustrates the SLB concept. The processing is of the form:

$$If \quad \frac{return_{main}}{return_{aux}} \geq k_{SLB} \frac{g_{main}}{g_{aux}} \quad then \quad keep; \ otherwise \ edit \ return \ , \qquad (8.1)$$

where $return_{main}$, $return_{aux}$, k_{SLB}, and g_{aux} are the main and auxiliary antenna returns, respectively, an SLB gain factor, and the main and auxiliary antenna voltage gains, respectively.

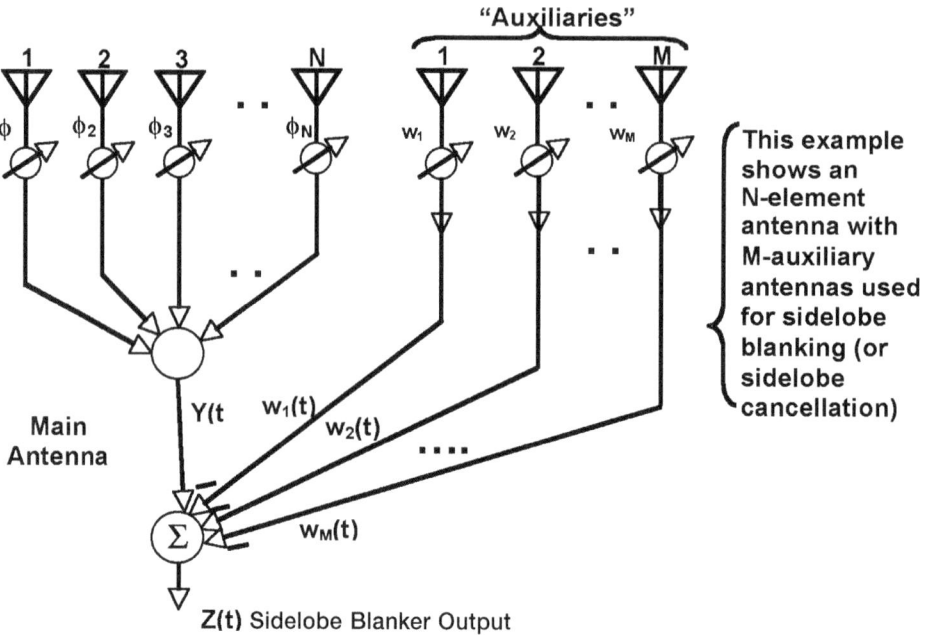

Figure 8.1 Basic Sidelobe Blanker Processing

8.3.2 Sidelobe Canceller

A basic sidelobe canceller (SLC) is depicted in Figure 8.2. SLCs also use auxiliary antennas. However, the SLC applies a weighted summation of the auxiliary antenna signals to that of the main antenna with the objective of cancelling an interference source.

These types of interference suppression can compute the complex auxiliary weight using either closed-loop (i.e., using feedback nulling) or open-loop (i.e., no feedback employed) techniques to estimate jammer powers and angles-of-arrival (AOA) to use for the SLC process. SLC performance in cancelling and interference source is dependent on the weight computation method as well as the jammer power, the associated RF bandwidth, and their angular direction of arrival (i.e., where in the antenna sidelobe structure they impinge upon the antenna).

For very narrowband interference (i.e., tones or sinusoids), an SLC can achieve near-perfect cancellation. However, for wideband interference, MSLCs (also re-

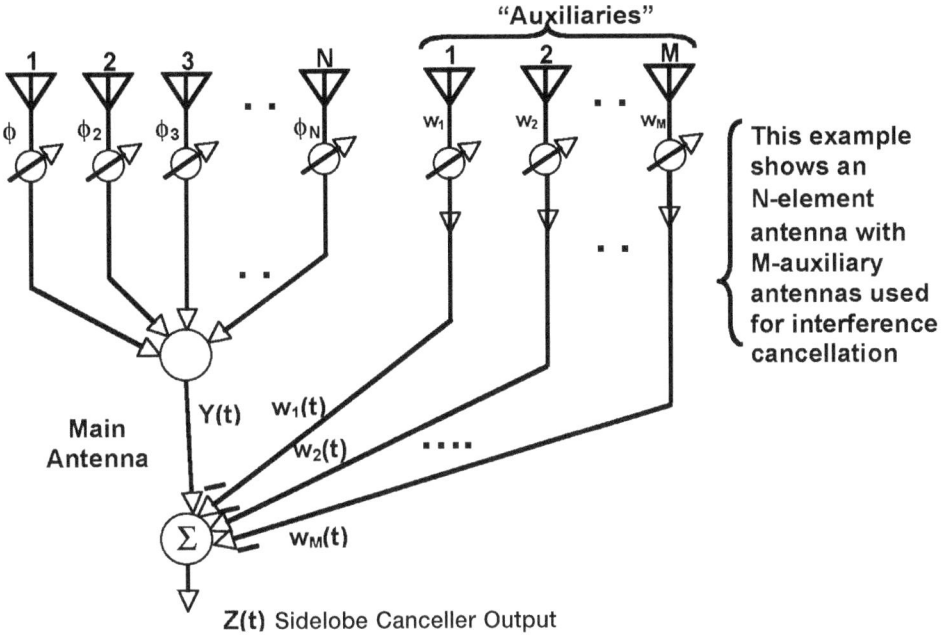

Figure 8.2   Basic Sidelobe Canceller Processing

ferred to as multiple degrees-of-freedom) can be required to cancel a single source.

The optimal weight estimate computations are shown in Figure 8.3 for a single SLC. When the optimal weight is used, the residual interference after cancellation is given by:

$$P_{min} = P_n \left[1 + \frac{P_n P_J G_m + P_J^2 G_m G_a}{\left(P_n + P_J G_a \right)^2} \right. \\ \left. + \left[\frac{\left(P_J^2 P_n G_m G_a + P_J^3 G_m G_a^2 \right)\left(1 - \rho^2(\tau) \right)}{\left(P_n + P_J G_a \right)^2} \right] \right]' \tag{8.2}$$

where P_n, P_J, G_m, G_a, and $\rho = 1$ are the noise and jammer powers, respectively, and the main and auxiliary antenna gains, respectively, and the correlation coefficient between the main and auxiliary channels. A simplifying case occurs:

$$w_{opt} = r_0(t)\, r^{-1}(t)$$

Optimal SLC weight

$$r_0(t) = E\{y_m(t)\,y_a^*(t)\}$$
$$r(t) = E\{y_a(t)\,y_a^*(t)\} = r^*(t)$$

Cross-correlation and auto-correlation

$$\hat{r}_0(t_N) = \frac{1}{N}\sum_{i=1}^{N} y_m(t_i)\,y_a^*(t_i) \longleftarrow$$

Estimates of cross-correlation and auto-correlation

Optimal weight estimate

$$\hat{r}(t_N) = \frac{1}{N}\sum_{i=1}^{N} y_a(t_i)\,y_a^*(t_i) \qquad then: \qquad \hat{w}_{opt}(t_N) = \hat{r}_0(t_N)\,\hat{r}^{-1}(t_N)$$

Figure 8.3 Optimal SLC Weight Computations

$$for \;\; P_J \gg P_n \;\; and \;\; \rho = 1 \qquad P_{min} \approx P_n\left[1 + \frac{G_m}{G_a}\right]. \tag{8.3}$$

8.3.3 Multiple Sidelobe Canceller

A multiple sidelobe canceller is similar to the SLC depicted in Figure 8.2, except that the optimal weight estimates are computed jointly. The derivation of the optimal weight vector is shown in Figure 8.4.

When the optimal weight vector is used, the residual interference is given by:

$$P_{Min} = P_{Main} - R_0^{\;H} R^{-1} R_0 = P_{Main} - R_0^H W_{OPT} \;\;, \tag{8.4}$$

where the optimal weight vector is computed as:

$$W_{OPT} = R^{-1} R_0 \;\;, \tag{8.5}$$

$$P_{res} = E\left\{r(t)r^*(t)\right\} = E\left\{y_0(t)y_0^*(t)\right\} - E\left\{Y^H(t)y_0(t)\right\}W$$
$$- W^H E\left\{Y(t)y_0^*(t)\right\} + W^H E\left\{Y(t)Y^H(t)\right\}W$$

$$P_{res} = P_{main} - R_0^H(t)W - W^H R_0(t) + W^H R(t)W \qquad R_0(t) = E\left\{Y(t)y_0^*(t)\right\}$$
$$R_0^H(t) = E\left\{Y^H(t)y_0(t)\right\}$$

Find W that minimizes the residual: $\qquad\qquad R(t) = E\left\{Y(t)Y^H(t)\right\}$

$$\frac{\partial P_{res}}{\partial W^H} = 0 = 0 - 2R_0(t) - 2R(t)W \qquad\qquad \Rightarrow \qquad W_{opt} = R^{-1}(t)R_0(t)$$

$$P_{min} = P_{main} - R_0^H R^{-1} R_0 - R_0^H R^{-1} R_0 + R_0^H R^{-1}RR^{-1} R_0$$
$$= P_{main} - R_0^H R^{-1} R_0 - R_0^H R^{-1} R_0 + R_0^H R^{-1} R_0$$
$$= P_{main} - R_0^H R^{-1} R_0 = P_{main} - R_0^H W_{opt}$$

Figure 8.4   Computation of Residual Interference after Cancellation

$$y_0(t) = n_0(t) - \sum_{k=1}^{N_J} g_m(\theta_{Jk})\, j_k(t)$$

$$y_n(t) = n_n(t) - \sum_{k=1}^{N_J} g_n(\theta_{Jk})\, j_k(t-\tau_{nk})\, e^{-j\omega\tau_{nk}} \qquad\qquad \tau_{nk} = \frac{d_n\,\sin\theta_{Jk}}{c}$$

$$P_{main} = E\left\{y_0(t)y_0^*(t)\right\} = P_n + \sum_{k=1}^{N_J} G_m(\theta_{Jk})\, P_{Jk} \qquad \textbf{(for uncorrelated jammers)}$$

Main and auxiliary correlation vector components:

$$r_{0n}(t) = E\left\{y_n(t)y_0^*(t)\right\} = \sum_{k=1}^{N_J} g_m(\theta_{Jk})\, g_n(\theta_{Jk})\, E\left\{j_k(t)\, j_k(t-\tau_{nk})\, e^{-j\omega\tau_{nk}}\right\}$$

$$= \sum_{k=1}^{N_J} g_m(\theta_{Jk})\, g_n(\theta_{Jk})\, P_{Jk}\, e^{-j\omega\tau_{nk}}\, \rho(\tau_{nk})$$

Aux cross-correlation matrix components (for equal main and aux channel noise powers):

$$r_{np}(t) = E\left\{y_n(t)y_p^*(t)\right\} = P_n\,\delta(n-p) + \sum_{k=1}^{N_J} g_n(\theta_{Jk})\, g_p(\theta_{Jk})\, P_{Jk}\, e^{-j\omega\left(\tau_{nk}-\tau_{pk}\right)}\, \rho(\tau_{nk}-\tau_{pk})$$

Figure 8.5   Computations of Cross-Correlation Vector and Covariance Matrix

and:

$$r_{no} = E\left\{y_n(t)y_o^*(t)\right\} = \sum_{k=1}^{N_J} g^*(\psi_k)\exp\left(-j\phi_{nk}\right)P_{J_K}\rho\left(\tau_{nk}\right) \qquad (8.6)$$

$$\begin{aligned} r_{nm} &= E\left\{y_n(t)\,y_m^*(t)\right\} = P_n\,\delta(n-m) \\ &+ \sum_{k=1}^{N_J} \exp\left[-j2\pi(d_n-d_m)f_0\sin\psi_k/c\right] \end{aligned} \qquad (8.7)$$

8.3.4   Adaptive Processing

Figure 8.6 shows three applications of adaptive processing. The three applications of adaptive processing cover the major types required by phased-array radars:

• Adaptive antenna array for interference cancellation

• Adaptive Doppler processing for clutter cancellation

• Adaptive array and Doppler processing for both.

The adaptive weight computations are very similar to those performed by MSLCs, except that instead of using auxiliary antennas separate from the main antenna, elements or subarrays of the main antenna are weighted to cancel interference or clutter. The adaptive weight solution is provided in Figure 8.7. An example of adaptive array performance in cancelling three jammers is illustrated in Figure 8.8 The sampled-matrix inversion (SMI) approach to computing the optimal weights is shown in Figure 8.9.

8.3.5   Digital Beam Forming

Digital beam forming allows more flexibility than adaptive arrays with fixed or hardwired beam formers. Here, digital receivers are placed immediately behind antenna elements or subarrays to supply multiple channels of digital data to a specialized signal processor referred to as DBF processing. Figure 8.10 depicts a DBF architecture with adaptive spatial and Doppler (or time-domain) control at

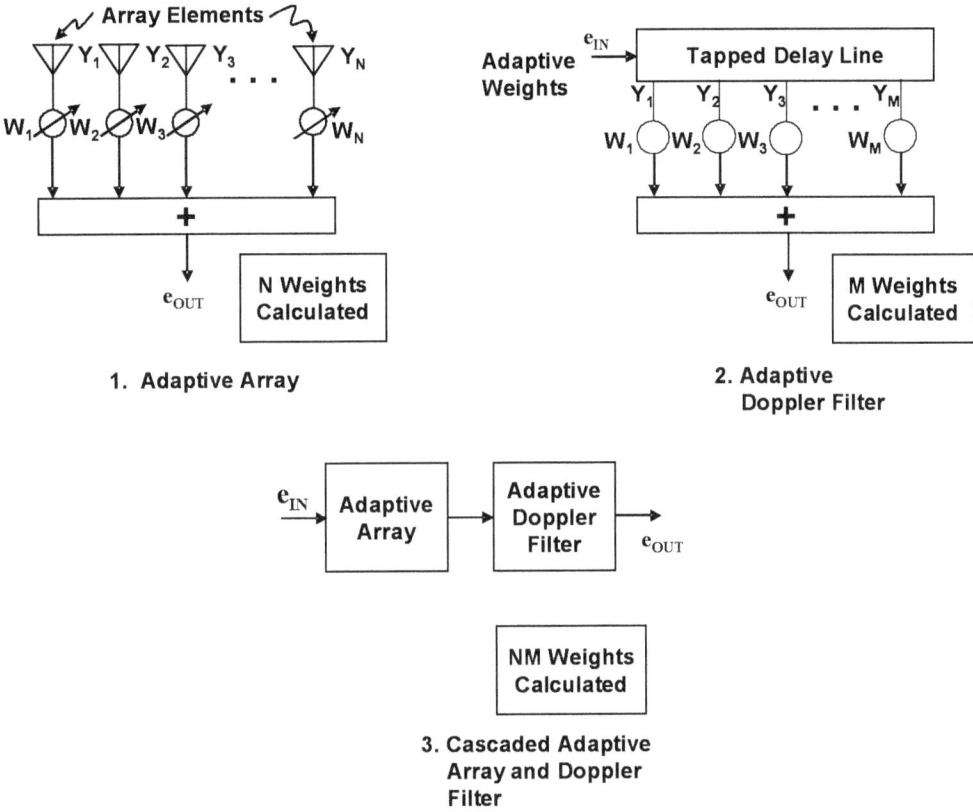

1. Adaptive Array

2. Adaptive Doppler Filter

3. Cascaded Adaptive Array and Doppler Filter

Figure 8.6   Three Adaptive Processor Examples

each subarray, which can cancel narrow and wideband interference and/or clutter.

Advantages of this DBF configuration are:

• Less hardware is required

  – Lower signal processing load than fully adaptive array

• Performance equal to fully adaptive if the number of subarrays ≥ interference degrees of freedom.

A disadvantage of subarray DBF is that the trade-off of hardware and processing versus degrees of freedom may not result in optimal or adequate interference reduction.

$$P_{OUT} = E\left\{\left|W^H Y\right|^2\right\} = E\left\{W^H Y Y^H W\right\} = W^H E\left\{Y Y^H\right\} W$$
$$= W^H R W \qquad where \qquad R = E\left\{Y Y^H\right\}$$

$$S_{OUT} = \left|W^H S\right|^2 = W^H S S^H W$$

$$SINR = \frac{S_{OUT}}{P_{OUT}} = \frac{W^H S S^H W}{W^H R W}$$

Maximize SINR subject to constraint : $\quad W^H R W = 1$

Define : $\quad F = W^H S S^H W - \lambda\left[W^H R W - 1\right]$

$$\frac{\partial F}{\partial W^H} = 0 = S S^H W - \lambda R W = W^H S S^H W - \lambda W^H R W$$

$$= 0 = W^H S S^H W - \lambda \qquad \Rightarrow \quad \lambda = W^H S S^H W$$

$$\frac{\partial F}{\partial W^H} = 0 = S S^H W - \lambda R W = S S^H W - \left(W^H S S^H W\right) R W$$

$$= 0 = S - W^H S R W \quad \Rightarrow \quad R W = \frac{1}{W^H S} S = g S$$

$$W_{OPT} = g R^{-1} S \qquad g \neq 0$$

Figure 8.7 Derivation of Adaptive Weight Vector

Figure 8.11 represents a similar DBF architecture at the element level. An advantage of this DBF architecture is that it provides the most degrees-of-freedom for cancelling narrow and wideband interference and/or clutter. However, except for small arrays, it can be prohibitively expensive to implement.

8.3.6   Frequency Agility and Hopping

Another means of countering interference that does not cover the entire radar operating band is to sense the frequency of the interference, and to operate at

Fully Adaptive Array
32 Array Elements (N=32)
Uniform Array Weighting
Jammers at -30, 20, and 50

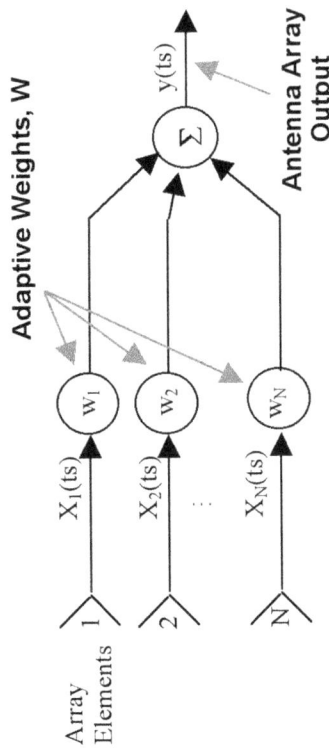

Adaptive Weights, W

$X_1(ts)$ w_1

$X_2(ts)$ w_2

$X_N(ts)$ w_N

Array Elements 1, 2, N

Σ $y(ts)$

Antenna Array Output

Assumptions:

• **Narrowband jammers**

• **Barrage noise**

Output SINR = $\dfrac{(w^H s s^H w)}{w^H R_{N+I} w}$

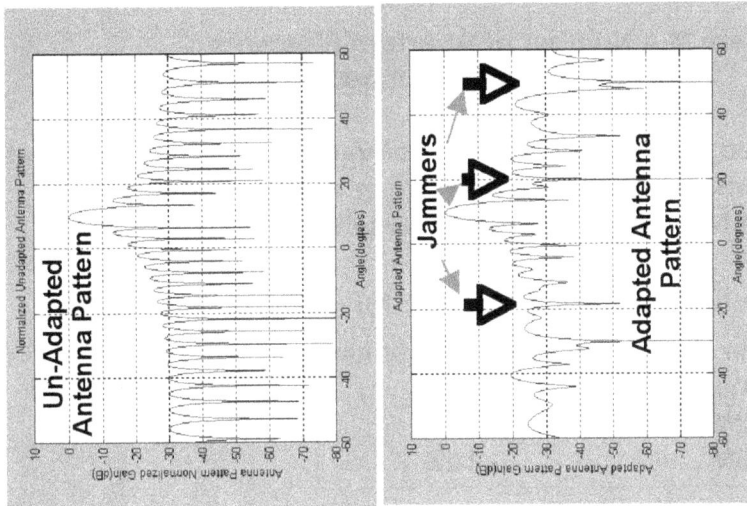

Normalized Unadapted Antenna Pattern

Un-Adapted Antenna Pattern

Adapted Antenna Pattern

Jammers

Adapted Antenna Pattern

Figure 8.8   Example of Adaptive Array cancelling Three Jammers

195

Must find R based on $\hat{R} = \dfrac{1}{N} \displaystyle\sum_{n=1}^{N} R(n)$

Where M = Number of Aperture Elements
N = Number of 'Snapshots' of Aperture

**Reed, Mallett, & Brennan Derived Expression for Deviation
From Optimum versus Number of Snapshots
and Number of Aperture Elements**

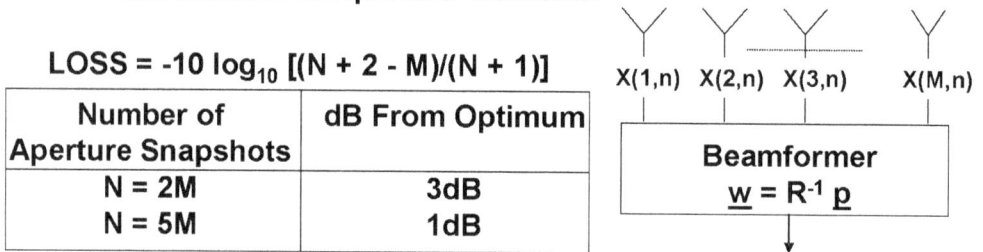

$LOSS = -10 \log_{10} [(N + 2 - M)/(N + 1)]$

Number of Aperture Snapshots	dB From Optimum
N = 2M	3dB
N = 5M	1dB

X(1,n) X(2,n) X(3,n)   X(M,n)

Beamformer
$\underline{w} = R^{-1}\,\underline{p}$

Figure 8.9   Sampled-Matrix Inversion Approach to Estimating Optimal Weight Vector

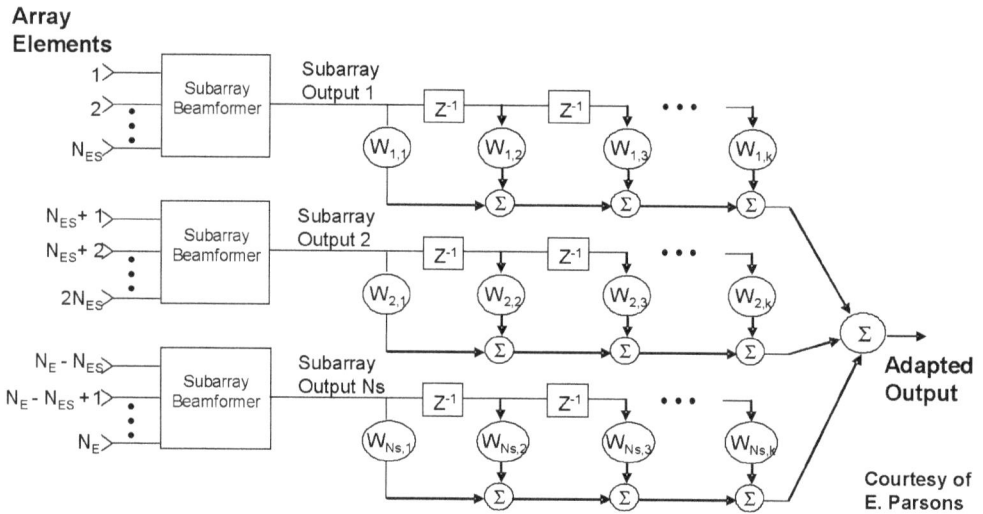

Figure 8.10   Space-Time DBF at the Subarray Level

another frequency. This is often the case for narrowband functions like search or track. Note that this presumes that the radar has alternative frequencies at which to operate. However, when very wideband radar waveforms are required for

Array
Elements

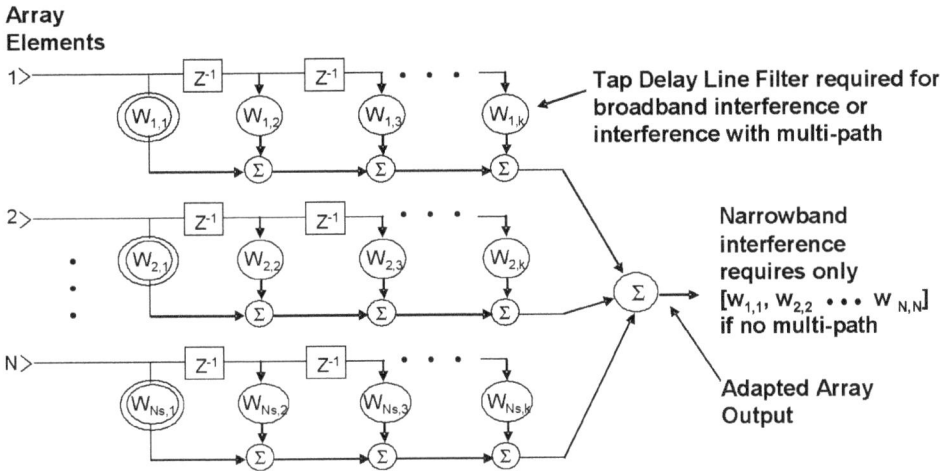

Figure 8.11   Space-Time DBF at the Element Level (courtesy of E. Parsons)

functions such as target classification there may not be alternative operating frequencies.

If responsive jammers are the interference source, then semi-random frequency hopping can be used to deny the jammer time to measure and produce false targets at the radar frequency since the radar's frequency is constantly changing in an unpredictable way.

8.3.7   Sector Blanking

When mainlobe interference sources are an issue, sector blanking is also an option. However, resorting to blanking on either transmit or receive (or both) sacrifices radar operation in those sectors. However, for limited angular sectors with fixed (or slowly moving) unintentional interferers, blanking can be a viable solution, since the small radar coverage lost may not be an issue.

8.4   PROBLEMS

8.4.1   Problem Statement

You are asked to assess the performance of a proposed adaptive processor for a technical proposal for a new radar development. Based on the technical requirements document (TRD) supplied with the Request for Proposal (RFP), the following electronic countermeasures (ECM) threat is specified as exhibiting:

$$\left(\frac{P_J}{P_S}\right)_{dB} = 35 \ \ \text{dB},\tag{1}$$

is a broadband jammer with greater than 10 MHz bandwidth and is situated at a slant range of 1000 km from the radar and at $\theta_J = +40$ degrees from antenna boresight.

The specified target parameters are:

RCS = –10 dBsm $\hfill$ (2)

Range = 500 km $\hfill$ (3)

$\theta_T = -20$ degrees $\hfill$ (4)

Target fluctuation model: Swerling I. $\hfill$ (5)

The proposed radar design provides the following capability:

SNR = 15 dB $\hfill$ (6)

RCS = –10 dBsm $\hfill$ (7)

Range = 500 km $\hfill$ (8)

Pulse length = 1 ms $\hfill$ (9)

Waveform bandwidth = 10 MHz. $\hfill$ (10)

The radar antenna is a linear phased array consisting of 10 elements spaced at $d = \lambda/2$. Amplitude and phase control are available at each antenna element. The weights are to be determined using an SMI)algorithm.

8.4.2 Task Description

The task consists of the following:

1. Calculate the optimum weight vector for the assumed target and jammer angles relative to antenna boresight.

2. Calculate the residual jammer-to-noise ratio (JNR) after applying the optimum weights as calculated in part 1.

3. Calculate the target probability of detection before and after applying the optimum weights.

4. If two equal power jammers are now located at $\theta_j = -40$ and $+40$ degrees relative to antenna boresight (with all other parameters the same), recalculate the adaptive weights, residual JNR, and target probabilities before and after adaptation.

5. Plot the adapted antenna patterns for the single-jammer and two-jammer cases. Compare the adapted antenna patterns to the un-adapted pattern. Comment on the differences.

8.4.3 Additional Information

The following information supplied by the proposal's radar systems engineers may be useful in completing the requested performance assessment:

(a) The covariance matrix elements and signal can be calculated from:

$$r_{nm} = P_n \delta(n-m) + \sum_{k=1}^{N_J} \exp[-2\pi j(n-m)(d/\lambda)\sin(\theta_k)]P_{Jk}\rho(\tau_{nk}-\tau_{mk}), \qquad (11)$$

where:

$$\tau_{nk} = (n-1)d \, \sin(\theta_k)/c, \qquad (12)$$

and:

$$S_n(t) = (SNR \; P_N)^{1/2} \exp[-j2\pi(n-1)(d/\lambda)\sin(\theta_T)]. \qquad (13)$$

For simplification, assume $\rho(\tau_{nk}-\tau_{mk}) = 1$, or perfect correlation.

(b) The output signal-to-interference-plus-noise ratio (SINR) is given by:

$$SINR = w^H S S^H w / w^H R \, w, \qquad (14)$$

where:

$$S = (SNR \, P_N)^{1/2}.$$ (15)

(c) The probability of detection for a Swerling I target is given by:

$$P_d = (P_{fa})^{\left(\frac{1}{1 + SINR}\right)},$$ (16)

where $SINR$ is a power ratio (i.e., not expressed in dB). The probability of false alarm (P_{fa}) is a design parameter. For the purposes of the performance assessment, assume $P_{fa} = 10^{-5}$, which is a reasonable value for surveillance.

8.5 REFERENCES

The following books are useful radar references on interference suppression.

[1] S. Haykin, *Adaptive Radar Signal Processing,* Wiley-Interscience, 2006

[2] S. Kay, *Modern Spectral Estimation: Theory and Application,* Prentice-Hall, 1999

[3] D. Manolakis, *Statistical and Adaptive Signal Processing,* Artech House, 2005

[4] S. L. Marple, *Digital Spectral Analysis with Applications,* Prentice-Hall, 1987

[5] R. A. Monzingo & T. M. Miller, *Introduction to Adaptive Arrays,* SciTech, 2003

[6] R. Nitzberg, *Radar Signal Processing and Adaptive Systems,* 2nd Edition, Artech House, 1999

[7] A. Oppenheim & R. Shafer, *Digital Signal Processing,* Prentice-Hall, 1975

[8] A. Papoulis, *Signal Analysis,* McGraw-Hill, 1977

[9] S. Haykin & A. Steinhardt, *Adaptive Radar Detection and Estimation,* Wiley, 1992

[10] J. V. Candy, *Signal Processing—The Modern Approach,* McGraw-Hill, 1988

9

Phased-Array Radar
Architectures

9.1 INTRODUCTION

This chapter examines some common phased-array radar (PAR) architectures used for many applications. Three general categories of architectures are described:

- Antenna-based
- Bandwidth-based
- Radar function-based.

The following sections cover these PAR architectures.

9.2 ANTENNA-BASED ARCHITECTURES

Four types of antenna-based PAR architectures are addressed in this section. They are:

- Full field-of-view (FFOV)
- Limited field-of-view (LFOV)

- Digital beam former (DBF)
- Mechanically steered phased-arrays.

9.2.1   Full Field-of-View Radars

The FFOV phased-array architecture is the most common form. It employs an antenna element spacing that ensures that there are no antenna grating lobes in real space. At a top level, this requires an average spacing of array elements $\leq \lambda/2$. Therefore, this is the most costly architecture in cost per unit area of active antenna aperture, with the advantage of the attendant maximum angular coverage.

Figure 9.1 illustrates the FFOV PAR architecture. It is characterized by phaseshifters at every array element, at least in its narrowband (NB) version. As discussed later, wideband (WB) phased arrays require the use of some degree of time-delay steering, usually implemented at the subarray level.

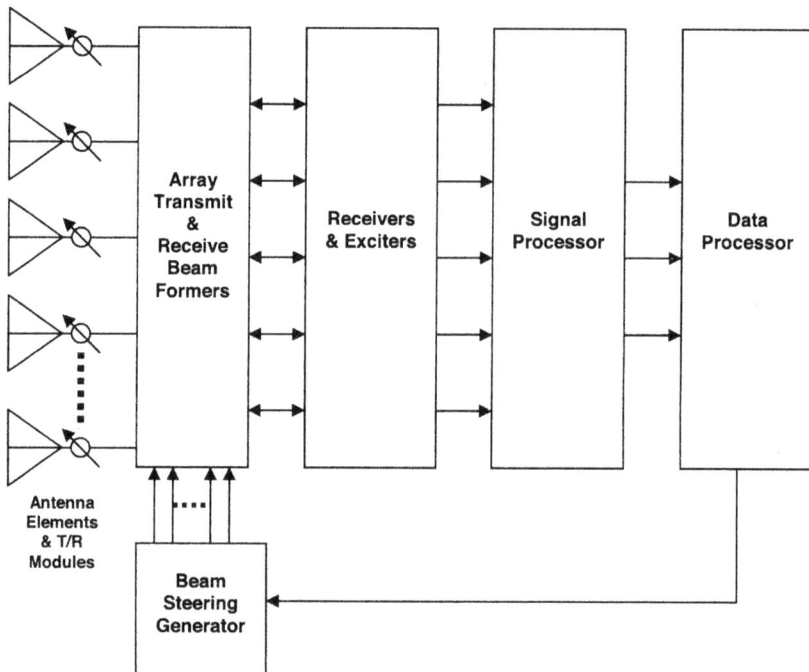

Figure 9.1   Full Field-of-View Phased-Array Architecture

As can be seen in the figure, this architecture consists of an array of antenna elements, each with a phase-shifter. The next components are the transmit and receive beam formers that synthesize the sum and azimuth and elevation monopulse patterns, as well as any auxiliary antenna patterns such as required for sidelobe blanker (SLB) or sidelobe canceller (SLC) functions. Each channel or port out of the receive beam former (RBF) has an associated receiver channel that converts the radio frequency (RF) at the operating frequency down to baseband and converts these analog signals to a digital format. Each of the digital data channels are next sent to the signal processor (SP), which performs matched filtering and detection processing. Finally, target return data from the SP are inputs to the data processor (DP), which performs search and track, as well as scheduling waveforms and controlling the hardware subsystems, including the beam steering generator (BSG), which translates the DP antenna steering commands to phase-shifter commands to each antenna element. On the transmit side, the exciters generate the waveforms at RF that are delivered to the transmit portion of the transmit/receive (T/R) modules, which consist of the transmit power amplifiers, receive amplifiers, and phase-shifters behind each antenna element.

As stated above, the FFOV PAR architecture is the most prevalent as it can provide the maximum angular coverage (typically ± 60 degrees in azimuth and/ or elevation angles) for radar functions such as search or surveillance and target tracking. Most of the PARs built today are of the active aperture or active array variety; that is, there are active transmitters behind each antenna element. Most T/R modules used today employ solid-state transistors for RF power amplification. However, many earlier radar systems employed passive arrays that were corporate-fed by one or more higher-powered transmitters, typically of the traveling-wave tube (TWT) type.

Most digital signal processing today is implemented by software residing in general or special purpose data processors. Often the signal and data processing software resides on a common signal/data processor (S/DP).

9.2.2 Limited Field-of-View Radars

Limited field-of-view (LFOV) radar architectures are a trade-off between maximum available electronic steering and the number of actively steered antenna elements. This architecture was developed specifically for large PARs operating at high frequencies such as X-band (10 GHz) that require less instantaneous angular coverage than available from FFOV radars, and, moreover, seek a lower-cost solution achieved via a reduction in the number of T/R modules. When more angular coverage is necessary over the long term (but not instantaneously), LFOV arrays can be mounted on an antenna pedestal or mount to provide a combination of mechanical and electronic steering. This radar architecture is attractive from a cost perspective.

Figure 9.2 illustrates the LFOV PAR architecture. On the surface it is very similar to the FFOV radar, except that the antenna consists of actively controlled super-elements that consist of the T/R modules that drive multiple passive antenna array elements. Since there are fewer active elements than for an FFOV radar of the same aperture size, and the super-element spacing is $\geq \lambda/2$, the LFOV antenna pattern exhibits grating lobes in real space. These must be minimized by careful array and subarray design, and the grating lobes must be actively controlled and monitored to prevent radiation safety hazards and to prevent other

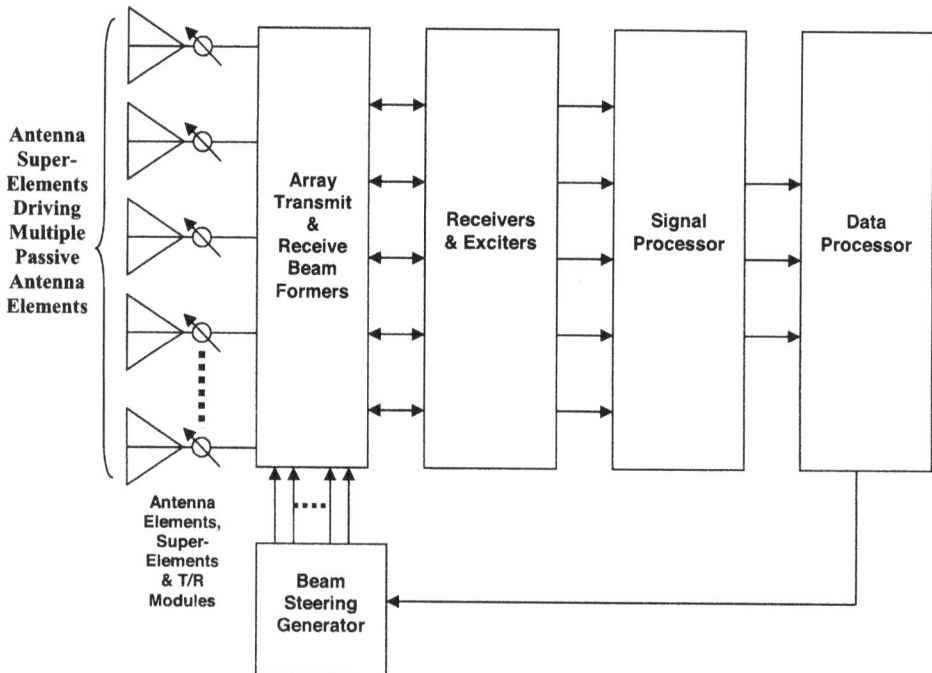

Figure 9.2 Limited Field-of-View Phased-Array Architecture

deleterious effects in clutter, interference, jamming, and multipath environments.

The remaining subsystems are essentially the same as for the FFOV architecture. The primary differences are in the transmit and receive beam formers and the BSG, which controls and steers a smaller number of super-elements than for an FFOV radar of the same aperture size. The LFOV does have approximately the same transmit and receive antenna gain as its similar-sized FFOV counterpart. However, due to the fewer active super-elements as compared with the number of active elements in the FFOV radar, the peak transmit power is reduced by the ratio of active LFOV super-elements to active FFOV elements.

9.2.3 Digital Beam Former Radars

In contrast to the "hard-wired" beam formers used in the previously described radar architectures, a relatively new type of radar architecture uses digital beam forming to digitally synthesize via software the antenna beams required by a particular radar's needs. This architecture is referred to here as a DBF radar architecture.

A DBF radar employs receivers at the front end of the radar to translate RF signals to baseband and convert analog signals to a digital format. The receivers can be behind each array element, each subarray, or some combination of subarrays (or equivalently at super subarrays). The result is a digital data stream from each array beam forming degree-of-freedom (DoF) that is available for processing by the signal processor or a DBF preprocessor preceding the SP to form whatever antenna beams are required. At a minimum it forms multiple sum beams and/or monopulse difference patterns (or the equivalent) for subsequent target detection and parameter estimation.

There are several advantages to the DBF architecture, including:

• Capability to form up to M independent sum beams to synthesize:
  – Simultaneous beams to reduce search occupancy
  – Stacked-beams for extended elevation search coverage

• Capability to form M independent sum pattern nulls

• Enhances the available dynamic range by a factor of up to M

- Supports implementation of single and multiple sidelobe cancellers
- Supports implementation of adaptive array processing
- Enables arbitrary antenna pattern shaping.

Figure 9.3 illustrates the basic DBF radar architecture. Contrasting Figure 9.3 to Figures 9.1 or 9.2 indicates that front-end hardwired beam formers are limited to those to form the array DoFs, that is, whether these are at the subarray of super-subarray level. If DBF is implemented at the element level, no hardwired beam former is required, and this component can be eliminated from the architecture shown in Figure 9.3.

Other changes for the DBF radar are the addition of the digital beam former processing. This can be implemented as a part of the SP or by a separate DBF pre-processor that precedes the SP. Also, the BSG is only required to steer the subarrays or super-subarrays depending on the DoFs implemented in antenna

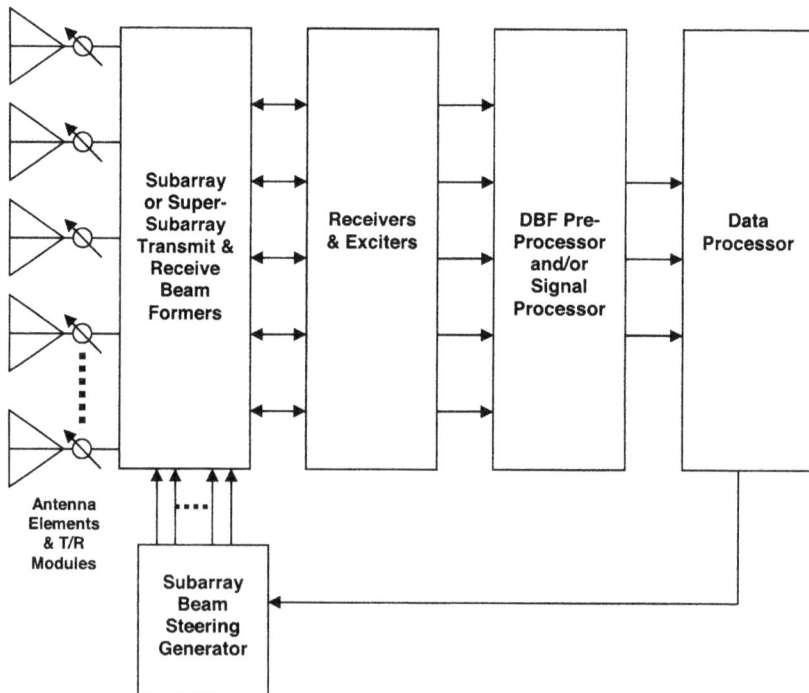

Figure 9.3   Digital Beam Former Phased-Array Architecture

array (i.e., whether DBF is implemented at the subarray or super-subarray level). The remaining subsystems and processing in the DBF architecture are similar to that of the hardwired beam former radars described in Sections 9.1 and 9.2.

9.2.4   Mechanically Steered Phased-Array Radars

These types of radar are a hybrid of a PAR and a mechanically scanned antenna. This architecture satisfies the requirement of greater than ± 60 degree (e.g., up to 360 degrees) coverage available when using a single-faced antenna array. These radars employ an antenna pedestal or mount to provide the mechanical steering capability. This architecture is appropriate when only a small instantaneous angular field-of-view (FOV) is required, but over a larger field-of-regard (FOR).

The basic mechanically steered PAR (MSPAR) architecture is illustrated in Figure 9.4. Effectively, the architecture is identical to that of the FFOV radar with the addition of an antenna pedestal or an antenna mount. Alternatively, when desired, the LFOV architecture shown in Figure 9.2 could be substituted for the FFOV architecture depicted in Figure 9.4.

9.3   BANDWIDTH-BASED ARCHITECTURES

Two basic types of bandwidth-based PAR architectures exist:

• Narrowband

• Wideband.

These two radar variants are described in the following sections.

9.3.1   Narrowband Radars

A precise definition of a narrowband PAR does not exist. However, it is generally understood that an NB radar employs a narrower bandwidth than a wideband radar. A WB radar is typically one that can instantaneously cover a bandwidth of at least 10 percent of its operating frequency. Using this definition, an NB radar is one that uses a bandwidth of less than 10% of its operating frequency.

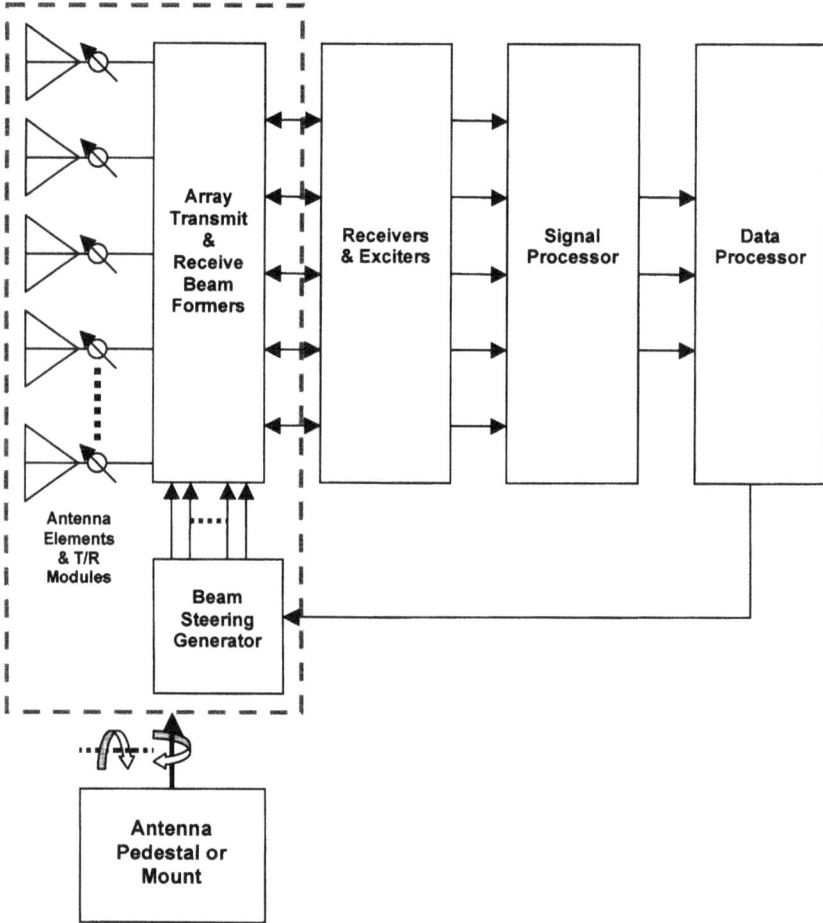

Figure 9.4   Mechanically Steered Phased-Array Architecture

From a radar architecture perspective, an NB radar can be of either the FFOV or LFOV variety, although they are typically the former. However, note that a radar that uses both NB and WB waveforms is in the WB radar class. A desirable feature of an NB radar is the ability to use phase-shift steering exclusively, that is, there is no need for time-delay steering. Under this requirement, the definition of the NB radar can be refined to be a radar that does *not* require time-delay steering. In general, this will constrain the radar instantaneous bandwidth of a NB radar to be much less than the 10% of operating frequency metric. Conversely, this will extend the definition of a WB radar to be one that uses more instantaneous bandwidth than its NB counterpart for a specified operating frequency. These definitions will be considered as adequate for describing and contrasting these classes of radars.

Typically, based on the above observations, NB radars will then support bandwidths of no more than 1% of their operating frequency. For example, an X-band radar that uses a 500 MHz would be classed a WB radar (i.e., 5% bandwidth), whereas one that employs 50 MHz would be considered an NB radar (i.e., 0.5% bandwidth).

Figure 9.5 illustrates a possible form of an NB radar; in this case it is also an FFOV radar. However, an LFOV radar (e.g., as illustrated in Figure 9.2) could also be substituted for Figure 9.5. The key discriminator is again the use of only phase-shift steering needed to electronically scan the antenna beam.

In addition to the electronic steering technique used, the bandwidth capability of most of the other subsystems in Figure 9.5 is likewise constrained. Specifically,

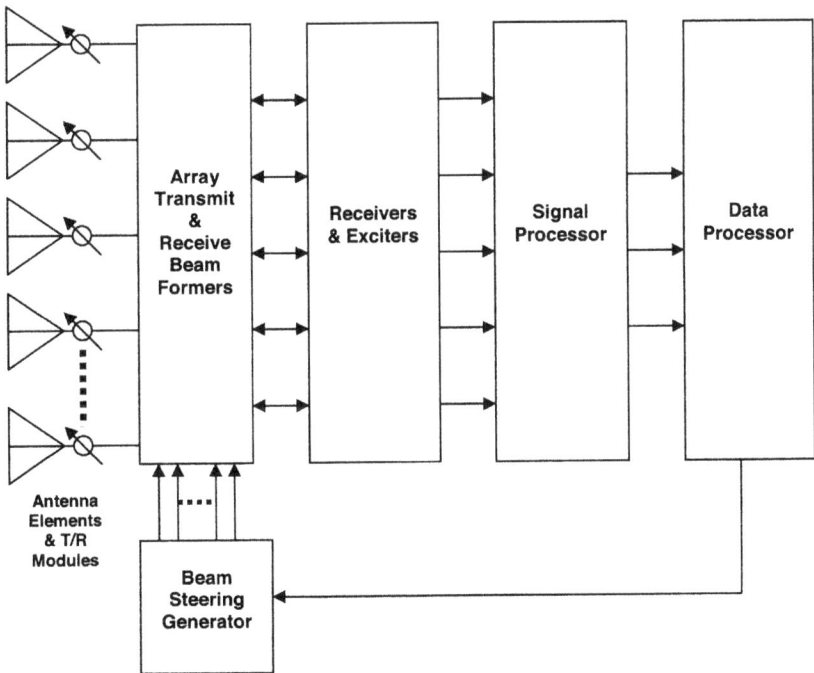

Figure 9.5 Narrowband Phased-Array Architecture

the antenna, beam formers, receiver, exciter, beam steering generator, and signal processing only need to support the narrower bandwidth. This impacts all active electronics in both transmit and receive paths.

Narrow bandwidth aspects include:

• NB waveform generation and RF up-conversion in the exciter

• NB down-conversion, intermediate frequency (IF) filters, and sample rates in the receiver

• NB power amplifiers in the T/R modules and subarray drivers

• Lower sampling rate processing, transform sizes, etc., in signal processor

• Narrow bandwidth modulation (e.g., linear frequency modulation; [LFM]) and associated matched filtering

• Phase-shift only computations in the BSG.

The narrowband nature of the resulting radar system and subsystems designs results, in general, in a radar of lower cost than its wide instantaneous bandwidth counterpart.

9.3.2 Wideband Radars

The alternative to the NB class of radars is the wideband class. As described in Section 9.3.1, this type of radar employs waveforms that occupy a larger percentage bandwidth than NB radars. Moreover, because of this difference, time-delay steering is required for WB radars. This is probably the most significant difference compared with NB radars.

Figure 9.6 illustrates the basic wideband radar architecture. The major changes relative to the front end of the NB radar architecture is the addition of time-delay steering, typically at the subarray level.

Other changes necessary to subsystems in the architecture include:

• WB waveform generation and RF up-conversion in the exciter

• WB down-conversion (and possible de-ramping or de-chirping), IF filters, and sample rates in the receiver

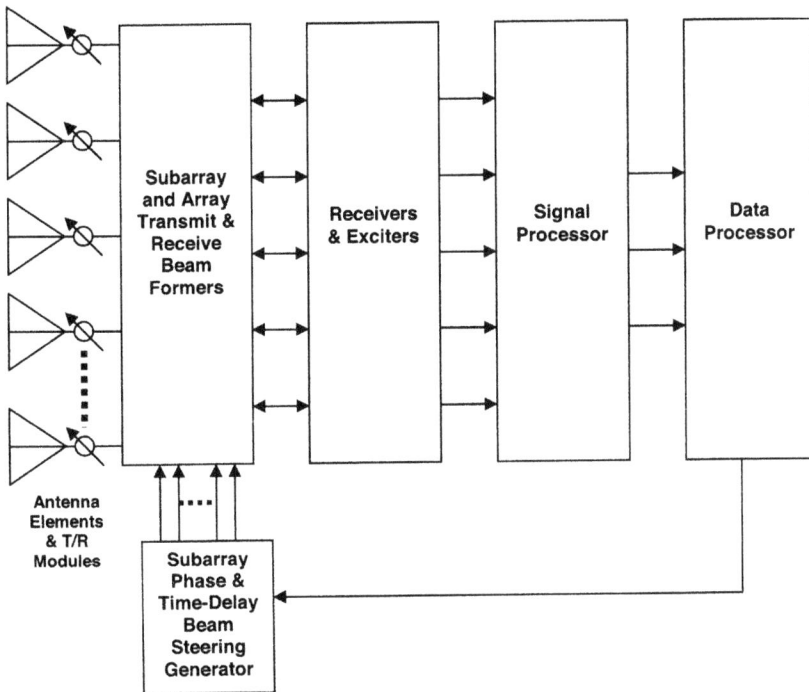

Figure 9.6   Wideband Phased-Array Architecture

- WB power amplifiers in the T/R modules and subarray drivers

- Higher sampling rate processing, transform sizes, etc., in signal processor

- Wide bandwidth modulation (e.g., LFM) and associated matched filtering (including both digital pulse compression and spectrum analysis or "stretch" processing as a function of receive window size)

- Both time-delay and phase-shifter computations in the BSG.

It should be noted that due to limitations in available A/D sampling rates and dynamic ranges, use of very wide bandwidths, for example, larger than 100 MHz, will preclude use of standard digital pulse compression techniques. The use of some form of "stretch" processing is the usual solution for very wideband operation. This requires some type of de-ramping or de-chirping (i.e., full or partial bandwidth) in the down-conversion process in the receiver, followed by spectral analysis in the signal processor.

9.4 FUNCTION-BASED RADARS

The last class of PAR architecture discussed is function-based radars. These include radars that perform:

• Search

• Track

• Target classification, discrimination, and identification (CDI)

• Missile illumination

• Multifunction.

The following sections describe these different PAR architectures.

9.4.1 Search Radars

The search class of radars is a very common and the simplest form besides that of missile illuminators. These radars are essentially NB radars that are primarily in the FFOV class, although they can also be implemented as mechanically steered LFOV radars. DBF radars can also be used as search radars due to their inherent multibeam capability, especially using "stacked" elevation beams and simultaneous multibeam operation to reduce search occupancy.

   Figure 9.7 illustrates an NB FFOV form of the search class of radars. Typically, very narrowband waveforms are employed, for example, usually in the range of several hundred kHz to about 1 MHz, often using LFM in clear environments. In clutter environments, narrowband moving target indicator (MTI), moving target detector (MTD), or pulse-Doppler waveforms are generally employed. Due to the very narrow bandwidth, all-range digital pulse compression is normally used. All subsystems are essentially the same as described for the NB FFOV radar architecture or for the DBF architecture when it is employed for search. Due to the usual large receive windows necessary to perform the search function; the signal processor may have to process very large fast Fourier transforms (FFTs) as part of the matched filtering when using digital pulse compression techniques. For clutter applications, some type of Doppler processing is also performed after pulse matched filtering in the SP.

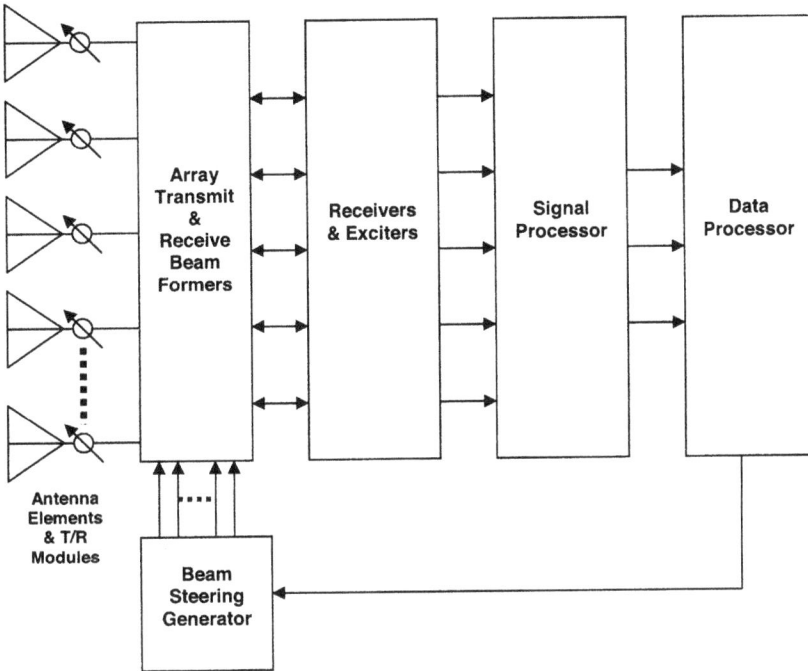

Figure 9.7 Search and Track Phased-Array Radar Architecture

9.4.2 Track Radars

Architecturally, the tracking radar class is nearly identical to that of the search radar described in Section 9.4.1, and therefore can likewise be implemented using NB FFOV, mechanically steered LFOV, or DBF class radar architectures.

Figure 9.7 also illustrates an NB FFOV form of the tracking class of radars. Typically, narrowband waveforms are employed, for example, usually in the range of 5 MHz to about 20 MHz, usually using LFM in clear environments. Again, in clutter environments, MTI, MTD, or pulse-Doppler waveforms are generally used. Due to the narrow bandwidth, digital pulse compression is normally used, except in cases when extremely large receive windows are required. Again, all subsystems are essentially the same as described for the NB FFOV radar architecture or for the DBF architecture when it is selected for tracking purposes.

Due to the usually small- to medium-sized receive windows necessary to perform the tracking function (as compared with search), the signal processor will usually have to process only medium-sized FFTs as part of the matched filtering when using digital pulse compression techniques. However, in cases where large receive windows are necessary based on mission requirements, digital pulse compression or stretch processing might be used depending on the specific waveform parameters and receive window sizes. In clutter environments, some form of Doppler processing is also performed after pulse-matched filtering. In addition to the normal detection processing used for search, tracking radars usually employ some form of parametric constant false alarm rate (CFAR) detection processing, and postdetection processing that includes monopulse for target angle measurement, and range and amplitude interpolation.

9.4.3   Classification, Discrimination, and Identification Radars

The CDI radar architectures are usually of the WB radar variety due to the requirement for wide bandwidth waveforms necessary to support target feature extraction. This is the case for both air and ballistic missile defense radars, although the specific waveform bandwidths and operating frequencies may differ for the two missions.

Figure 9.8 illustrates the basic CDI radar architecture, which is identical in structure to the WB radar class.

Key attributes of the CDI radar subsystems in the architecture shown include:

• WB antenna elements and T/R modules

• WB waveform generation and RF up-conversion in the exciter

• WB down-conversion (and possible de-ramping or de-chirping), IF filters, and sample rates in the receiver

• WB power amplifiers in the T/R modules and subarray drivers

• Higher sampling rate processing, transform sizes, etc., in signal processor

• Wide bandwidth modulation (e.g., LFM) and associated matched filtering (including both digital pulse compression and spectrum analysis or "stretch" processing as a function of receive window size)

• Both time-delay and phase-shifter computations in the BSG.

The prevalent use of some form of "stretch" processing is the case for the necessary very wideband operation. As in the case of WB radars, this requires some

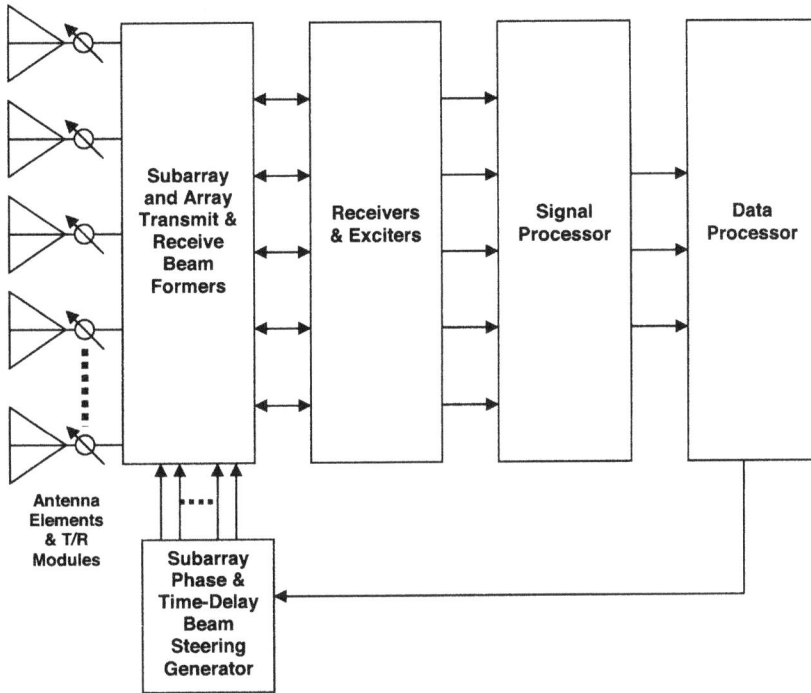

Figure 9.8 CDI Phased-Array Radar Architecture

type of de-ramping or de-chirping (i.e., full or partial bandwidth) in the down-conversion process in the receiver, followed by spectral analysis in the signal processor.

Due to the usually very small to small receive windows necessary to perform the CDI function (compared with tracking); the signal processor will usually have to process only medium-sized FFTs as part of the matched filtering when using stretch processing. In addition to the detection and postdetection processing required for tracking radars, CDI radars require target feature extraction, as well as feature conditioning algorithms.

9.4.4   Missile Illumination Radars

Missile illuminators are a special-case radar function used to support semi-active RF homing type interceptors in fire control applications. Phased-array architectures that provide this capability are usually of the NB FFOV type. Figure 9.5 is representative of a missile-illuminator radar architecture.

9.4.5   Multifunction Radars

This class of radar refers to those radars that support multiple radar functions. Typical variants include:

• Search and track

• Search, track, and identification (for air defense fire control)

• Search, track, identification, and illumination (for air defense fire control)

• Search, track, and classification and discrimination (for ballistic missile defense fire control).

The appropriate PAR architecture is usually the one that provides the highest level of functionality or performance required, often defined by the necessary waveform bandwidth. For example, the first case above might be an NB FFOV radar since only narrowband search and tracking are required, whereas the last one would most likely be a WB FFOV, WB LFOV, or mechanically steered WB LFOV radar since it is driven by the CDI function's requirements for wideband target classification and discrimination.

9.5   SCALABLE RADAR ARCHITECTURES

The concept of developing scalable radar architectures and the associated hardware and software architectures is described in the following subsections. Some notional system-level architecture constructs are introduced, and, from them, some candidate approaches to achieve scalability at the subsystem levels via the definition of candidate architectural building blocks.

9.5.1   Scalable Architecture Objectives

The objectives for developing scalable radar architectures are several-fold, with a few listed below:

- Ability to construct radars of varying sizes, for different missions and requirements using "radar building blocks"

- Radar building blocks would be used to synthesize all radar hardware and software subsystems

- Building blocks would constitute "product lines" as opposed to radars or complete subsystems being the product lines

- Building blocks, once developed, would be off-the-shelf with little effort required for documentation, testing, etc.

- Building blocks would have a minimum (ideally zero) of tailorability

- The performance and costs of building blocks would be known and stable

- An adequate number of hardware and software building blocks would be available to synthesize any radar, regardless of mission, size, performance, etc.

This list is incomplete but should give an indication of the goals of such an approach. Obviously, no one would question the advantages of having such architectural "widgets" from many perspectives, including design, implementation, performance, cost, reliability, extensibility, and maintainability.

9.5.2   Scalable Architecture Components

Given the objectives list in Section 9.5.2, some thought is required to understand what minimum set of architecture "components" might form a basis for achieving scalable radar architectures. It is instructive to first explore some of the commonly employed radar architectures used for existing missions and radar applications.

Five types of phased-array radar architectures are addressed in this section. They are:

- Full field-of-view

- Limited field-of-view

- Digital beam former

- Mechanically steered phased-arrays

- Wideband radars.

Figures 9.9 through 9.13 illustrate these basic radar architectures at the block diagram level and are repeated here from Section 9.2.

As can be seen in these block diagrams, at this high level of abstraction, there is a great deal of commonality with the five phased-array radar architectures shown. In order to better characterize distinct radar architecture elements, some fundamental hardware items are listed below:

• Antenna elements

• Antenna super-elements

• T/R modules

• Antenna array structures

• Antenna subarray structures

• Multiple-element structures

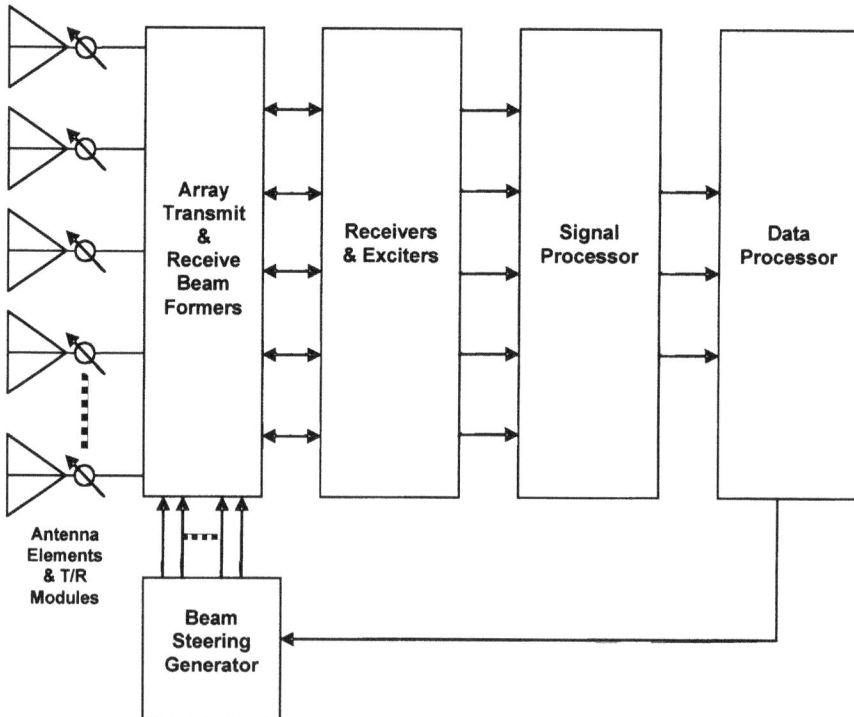

Figure 9.9   FFOV Phased-Array Radar Architecture

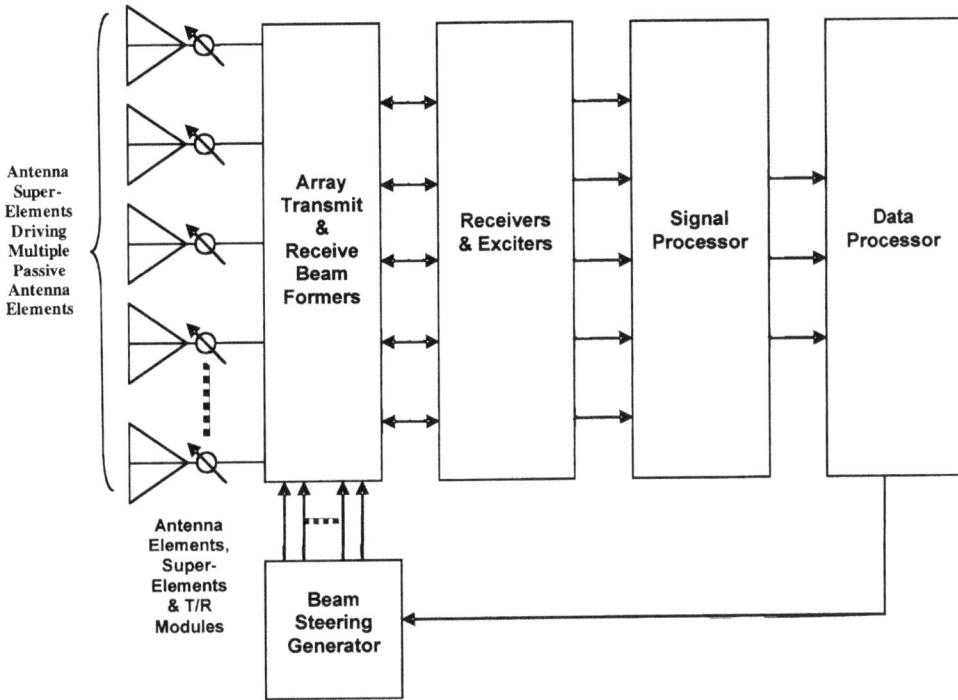

Figure 9.10   LFOV Phased-Array Radar Architecture

- Phase-shifters
- Time-delay units
- Array transmit and receive beamformers
- Subarray or super-subarray beamformers
- Beam steering generator
- Subarray beam steering generator
- Subarray phase and time-delay steering generator
- Narrowband receivers
- Wideband receivers
- Narrowband exciters
- Wideband exciters

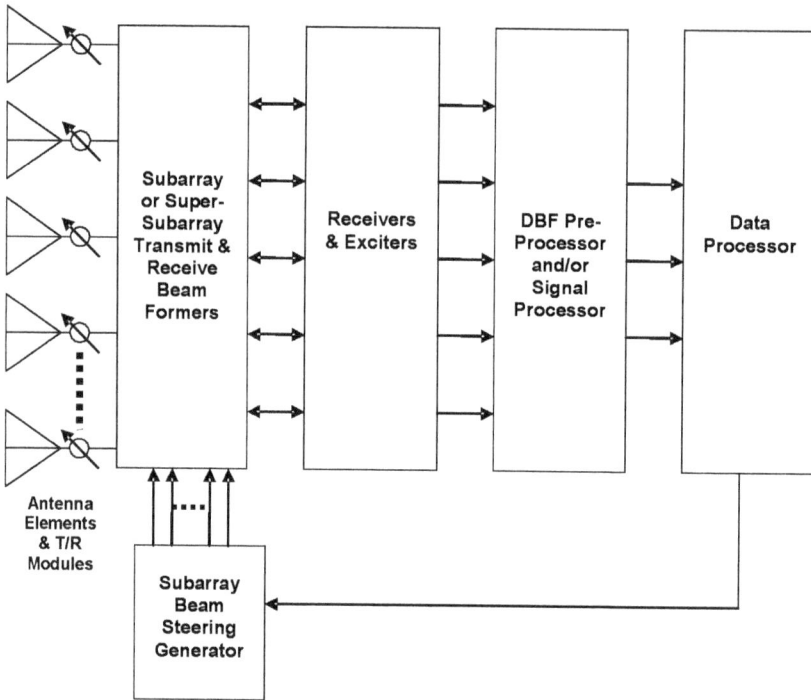

Figure 9.11   DBF Phased-Array Radar Architecture

- Signal processor
- Digital beam former processing
- Antenna pedestal or mount
- Antenna pedestal or mount controller
- Data processor
- Physical enclosures
- Power supplies and conditioning
- Cooling system.

This list, while incomplete, should give an idea of some of the hardware functional items necessary to implement many phased-array radars.

Now, within the data and signal processors, there are a number of software functional items, including:

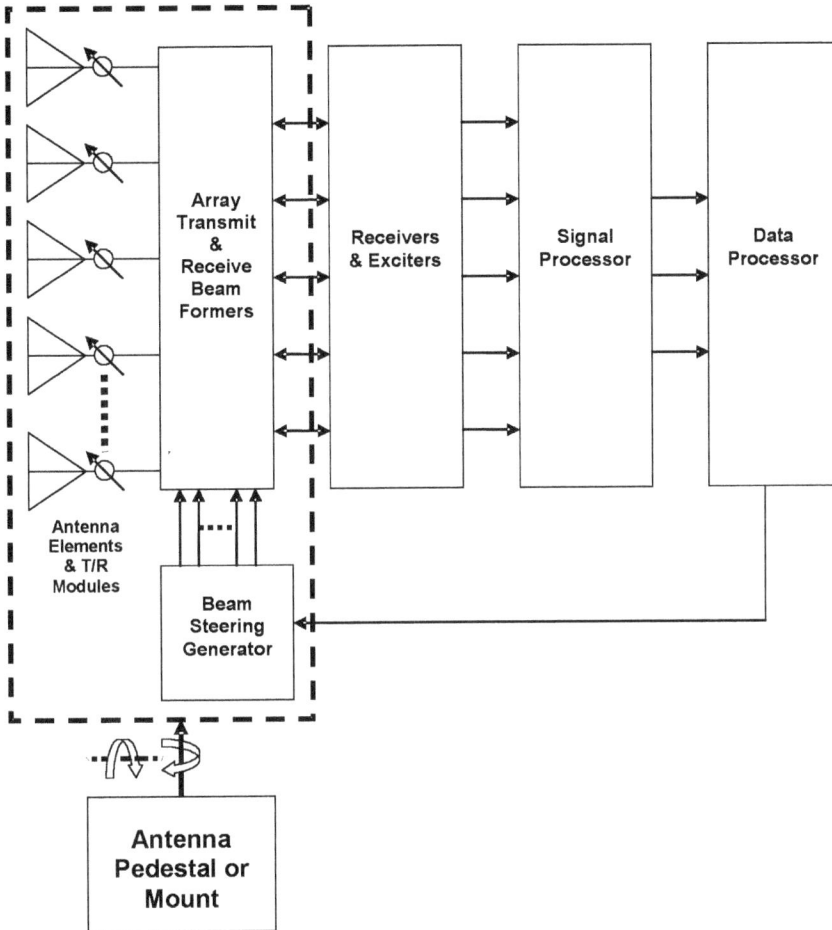

Figure 9.12 Mechanically Steered Phased-Array Radar Architecture

- Resource manager
- Radar scheduler
- Commands generation
- Returns processing
- Search processing
- Track processing
- Classification, discrimination, and identification processing

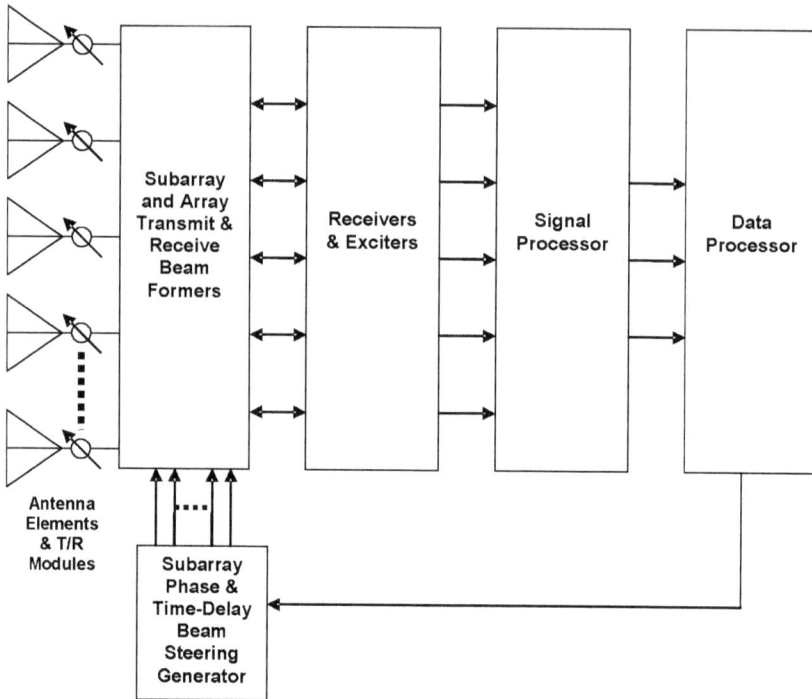

Figure 9.13   Wideband Phased-Array Radar Architecture

- Interceptor support
- Antenna pedestal/mount control
- Inertial navigation system/global positioning system (INS/GPS) processing
- Coordinate transformations
- Waveform matched filtering
- Detection processing (e.g., noise, CFAR)
- Postdetection processing (interpolation, peak detection, monopulse)
- Data recording
- Fault detection and fault isolation
- Calibration and alignment
- Built-in test

• Manual operation

• Digital simulation

• Hardware-in-the-loop simulation

• Scenario generation

• Operator displays

• Operator controls

• External communications

• Data report generation

• Pre-mission data generation.

Again, this list is not complete, but should give a good idea of the software processing "widgets" necessary for various types of phased-array radar applications.

9.5.3 Candidate Building Blocks for Scalable Radar Architectures

In order to define a candidate set of radar architectural building blocks, some basic rules must first be established. Obviously, since the goal is scalability with the objectives established in Section 9.5.2, the necessary rules are in the form of guidelines and constraints, including:

• Building blocks should include as much functionality as possible (within reason) that can be considered to be a fundamental radar element

• Building blocks can include both hardware and software functions

• Building blocks should be defined so as to minimize external interfaces and maximize internal interfaces

• Similarly, functions with high-interdependencies should be bundled together

• A building block may exist in a number of versions (e.g., based on operating frequency)

• Building blocks should have a minimum of tailorability.

The following section suggests some candidate building blocks.

9.5.3.1 Building Blocks

FFOV Subarray Block.   This building block consists of a complete FFOV subarray, including an exciter and multiple receiver channels, phase-shifters for each element and T/R module, subarray beam steering generator, AC-to-DC power conversion, built-in test functionality, power distribution, cooling, and a modular physical structure, and external interfaces for control inputs from a data processor, multiple channels of digital baseband data output, power inputs, and cooling intake/output. Figure 9.14 is a simplified block diagram of this building block.

The intent is that any number of these FFOV subarray blocks could be combined to achieve the desired antenna aperture size. The element spacing would

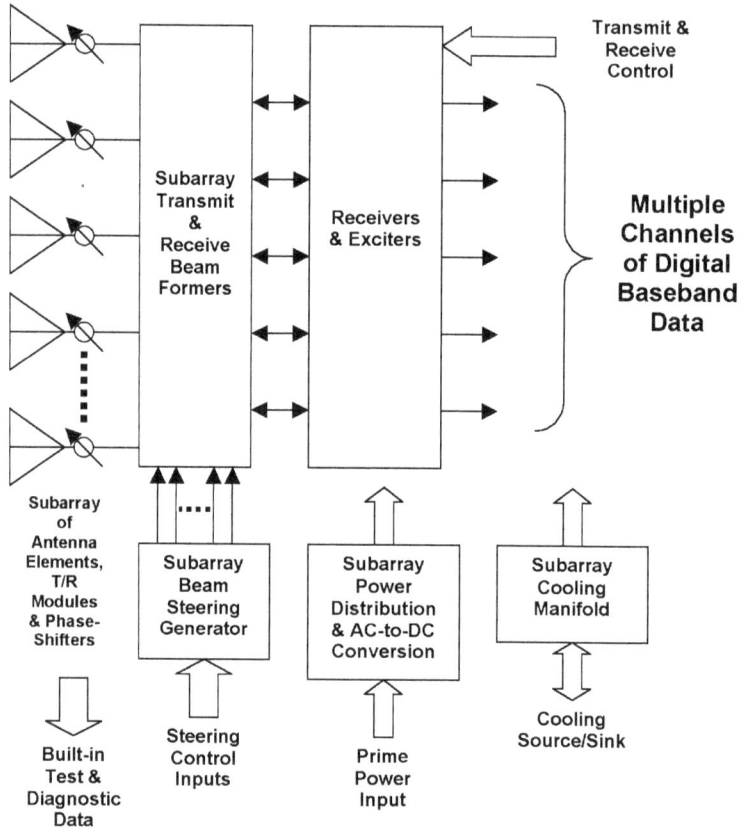

Figure 9.14   FFOV Subarray Block Functional Diagram

be a standard triangular grid that allows ± 60 degrees of azimuth and elevation scan without grating lobes. These blocks would be available at ultra-high frequency (UHF), L-band, S-band, C-band, X-band, and so on to facilitate modular construction of any size antenna aperture. The mechanical structure would be such as to allow building blocks to interlock and also allow ease of connection of input and output signals, power, and cooling for the overall antenna aperture.

The physical size of the subarray would be driven by the minimum size necessary to support wideband operation (as a function of operating band limitations) using a single time-delay unit (physical or otherwise) at each building block. This would enable the FFOV subarray block to also be a fundamental element of a wideband FFOV subarray block with wideband exciters/receivers, and time-delay applied at the building block level. This indicates the need for both narrowband and wideband variants of the FFOV subarray block.

LFOV Subarray Block. Figure 9.15 is the LFOV version of the FFOV subarray block. The key difference is the use of super-elements or horns in place of antenna elements, with larger spacing than required for FFOV operation. Again, both narrowband and wideband variants of the LFOV subarray block are required.

Signal/Data Processor Block. This building block would be a data processor with the adequate throughput, memory, and I/O capability to support signal and data processing software in a single unit. The processor would consist of a number of multiple servers (e.g., blades) that could be scaled easily within reason to increase computational capacity depending on the resident software needs.

Signal Processing Software Block. This software item would incorporate a number of the functions identified in Section 9.5.3. A simple block diagram of this building block appears in Figure 9.16.

Radar Hardware Control Software Block. Figure 9.17 represents a software block that provides radar hardware control. This block is effectively the software that interfaces with the FFOV or LFOV subarray blocks to provide transmit and receive action commands for waveform generation, transmission, reception, and the associated beam steers. In addition, this block performs the transmit beam

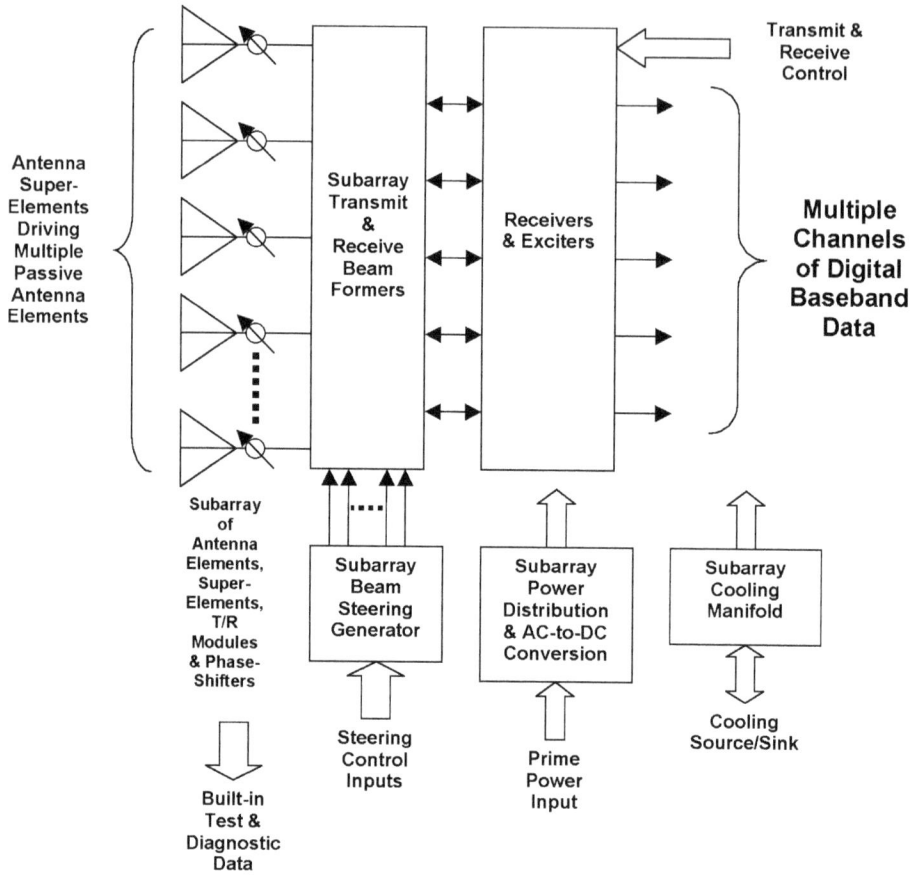

Figure 9.15   LFOV Subarray Block Functional Diagram

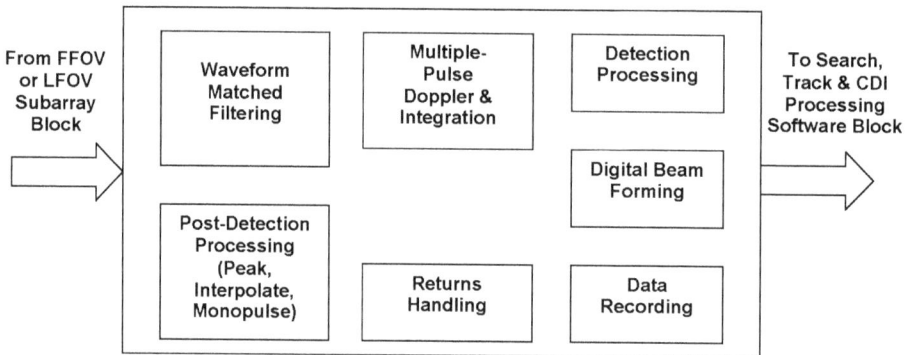

Figure 9.16   Signal Processing Software Block

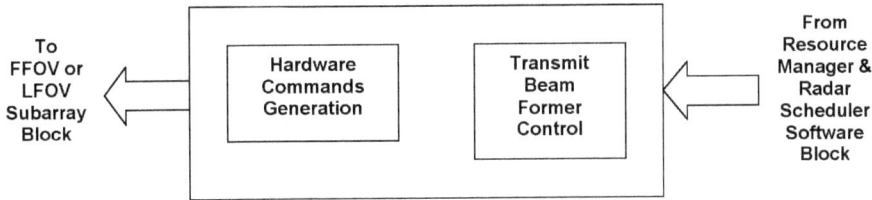

Figure 9.17 Radar Hardware Control Software Block

forming control. It is this function that controls partitioning of an antenna array and formation of multiple beams on transmit and implements time-delay steering for wideband transmit operation.

Resource Manager and Radar Scheduler Software Block. This software block represents the "brain" of the radar architecture. It consists of a resource manager that allocates radar duty and timeline occupancy to radar tasks, and a radar scheduler that assigns transmit and receive actions to the radar timeline. This block is depicted in Figure 9.18. There will be several variants of this block, for example, for:

• Ballistic missile early warning

• Air defense fire control

• Ballistic missile defense fire control

• Ship self-defense,

due to the very distinct requirements for functionality, latency, and performance of each of these radar applications.

Search, Track, and CDI Software Block. Figure 9.19 depicts the search, track, and CDI software block. This block entails the search processing, tracking logic (data

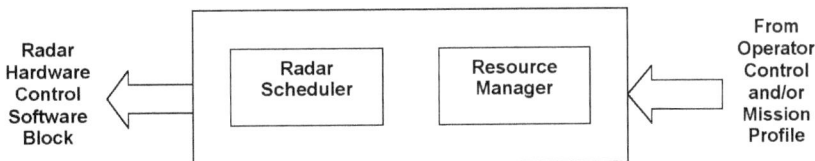

Figure 9.18 Resource Manager and Radar Scheduler Software Block

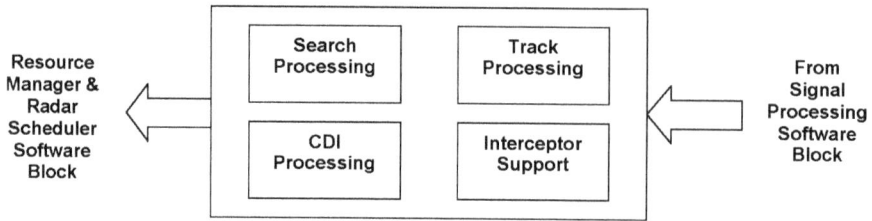

Figure 9.19 Search, Track, and CDI Software Block

association, track filters, track file maintenance), and the CDI algorithms necessary for radar applications. Also included are the interceptor support functions. Since this is closely associated with track-related processing and data, it is bundled into this software block. As identified under "Resource Manager and Radar Scheduler Software Block," at least four variants are required for different radar applications.

Antenna Pedestal and Pedestal Control Block. This hardware block includes both the antenna pedestal or antenna mount, and the antenna servo control system (ASCS) that controls the mechanical steering of the pedestal or mount. A simple block diagram is shown in Figure 9.20.

Power System Block. This hardware block consists of the power generation, power conversion, and power conditioning needed to supply necessary power at required voltages to all radar hardware blocks.

Cooling System Block. This hardware block consists of the cooling source (e.g., chiller, heat exchanger), cooling medium (e.g., water, glycol solution), and transmission method (e.g., pumps, hoses) needed to supply the cooling for all radar hardware blocks.

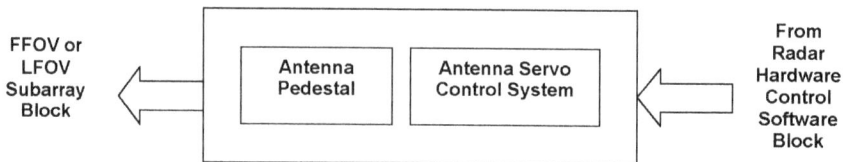

Figure 9.20 Antenna Pedestal and Pedestal Control Hardware Block

FD/FI, Calibration, and Alignment Software Block. This software block includes the radar system-level fault detection, fault isolation, calibration and alignment functions that collect and assess radar health data, and compute calibration and alignment data based on pilot-pulse processing and similar calibration functions.

Operator Displays and Controls Hardware/Software Block. This hardware/software block consists of the operator controls and displays needed to operate the radar system.

Scenario Generation and Simulation Software Block. This software block generates target and environment (e.g., ship motion, clutter, jamming, decoys) scenarios, and controls and implements digital simulation and hardware-in-the-loop simulation or analog simulation functions. This block is depicted in Figure 9.21.

Pre- and Post-Processing Software Block. This software block includes a tool necessary to create pre-mission data (e.g., waveform matched filter replicas, mission profiles) and to process, reduce, and compile data reports needed for testing and performance assessment.

External Communications Hardware/Software Block. This hardware/software block provides the interface to all external systems and handles all necessary data translation and formatting.

Utilities and Services Software Block. This software block would include all utilities and services, such as INS/GPS processing, coordinate transformations, time and date, and other similar types of functions needed to support overall radar operation.

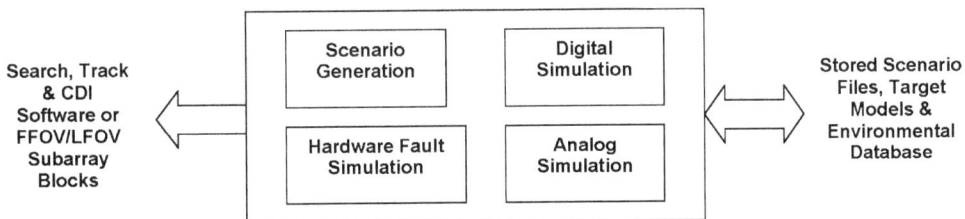

Figure 9.21 Scenario Generation and Simulation Software Block

9.5.4 Examples of Radars Synthesized from Architecture Building Blocks

In order to demonstrate the use of the scalable architecture concept, two examples of radar systems are synthesized using the hardware and software building blocks defined in Section 9.4.

9.5.4.1 Wideband FFOV Radar. Figure 9.22 depicts a WB fire control radar synthesized from architectural building blocks.

9.5.4.2 Mechanically Scanned LFOV Radar. Figure 9.23 depicts a mechanically scanned LFOV radar synthesized from architectural building blocks.

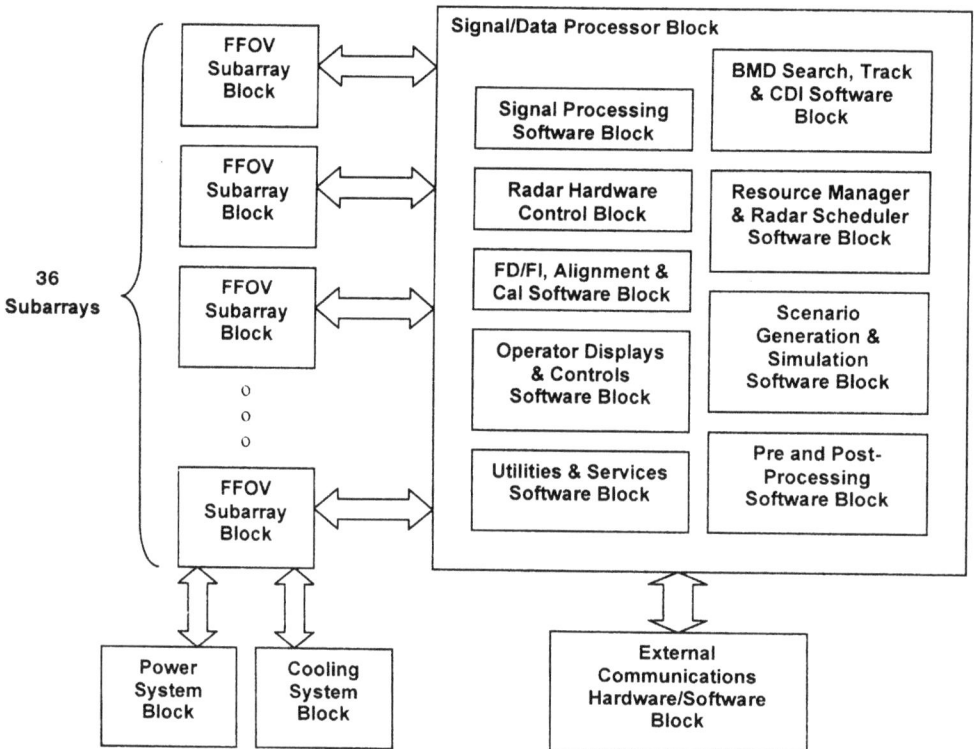

Figure 9.22 Wideband FFOV Radar Synthesized from Building Blocks

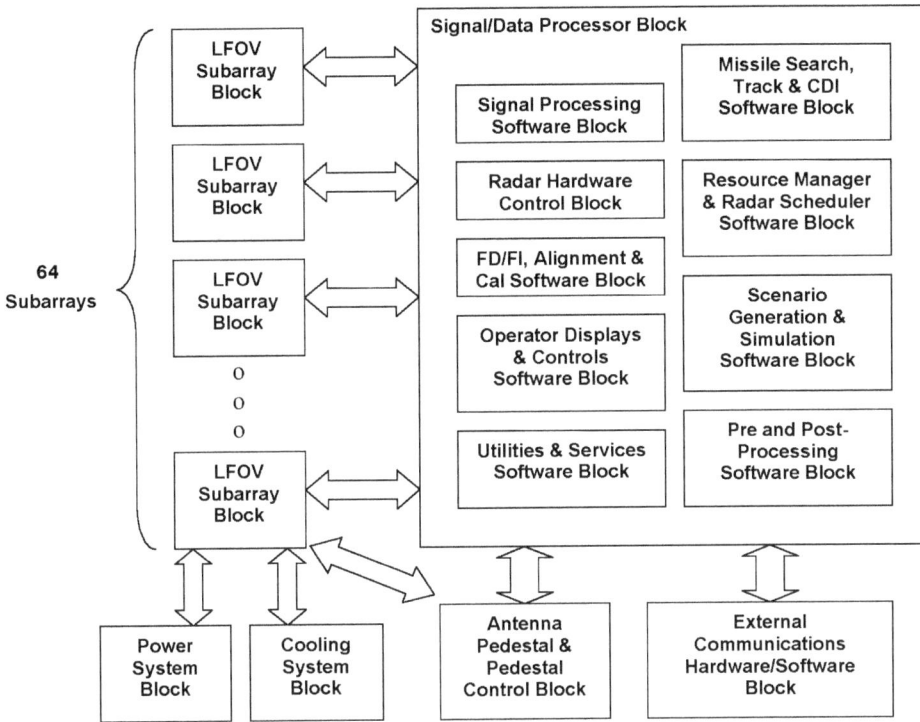

Figure 9.23 Mechanically Scanned LFOV Radar Synthesized from Building Blocks

10

Fundamental Radar Design Trade-Offs

10.1 INTRODUCTION

There are a number of trade-offs to be made in the design of a radar for any application. Some major categories and specific types of trade studies that are performed include:

- Operating frequency selection:
  - For search and track
  - For target classification
  - For operation in clutter, jamming and chaff
  - For early warning applications
  - For air defense applications
  - For missile defense applications
  - For surface target search and track
- Waveform selection:
  - In the clear:
    Search
    Track
    Target classification and identification

  - In clutter:
    Search
    Track
    Target classification and identification
  - In chaff:
    Search
    Track

- Radar coverage:

  - Range
  - Angle
  - Doppler

- Receiver operating characteristic design:

  - Target fluctuation types
  - Probabilities of false alarm and detection
  - Coherent integration (CI) and non-coherent integration (NCI)
  - Pulse repetition frequency (PRF)

- Search design:

  - Target types, fluctuation models, and dynamics
  - Search fence versus volumetric search
  - Coherent and non-coherent integration
  - Cumulative probability approach (e.g., binary, M-of-N)

- Tracking architecture and parameter selection:

  - Target types and dynamics
  - Data association algorithm selection
  - Track filter and model selection

- Target classification selection:

  - Air targets
  - Ballistic missile targets
  - Ship and vehicle targets.

This chapter discusses these trade studies and provides an overview of the associated objectives and approaches of each type.

10.2 OPERATING FREQUENCY SELECTION

In most cases, this is the first trade study performed in the design of a radar system. There are many factors that influence the selection of frequency, including:

• Radar functions to be performed

• Operating environment

• Radar mission

• Target types to be handled.

The types of radar and the functions they perform typically have the strongest influence on frequency selection. For the most part the selection of operating frequency can be analytically driven by exercising the appropriate form of the radar range equation (RRE). The following sections explore this approach.

10.2.1 Volume Search

Chapter 1 provides different forms of the RRE developed to support the key radar functions, such as volume search, horizon fence search, track sensitivity, and track accuracy. Referring to Chapter 1, the volume search form of the RRE is:

$$SNR = \frac{\sigma\, T_{SC}}{\left(4\pi\right) k T_s\, R^4\, \Psi L_t\, L_r}\, P_{AVE}\, A_r,$$
(10.1)

where the parameters σ, T_{SC}, T_s, R, Ψ, L, P_{AVE}, and A_r are the radar cross section (RCS), search scan or frame time, system noise temperature, target slant range, search solid angle, transmit and receive losses, average transmitter power, and receive aperture, respectively.

As noted in Chapter 1, there is no explicit frequency dependence indicated in equation (10.1). Therefore, aside from the system noise figure and losses, search sensitivity (SNR) is only a function of the average power-aperture product. This suggests that two topics be considered in selecting the radar operating frequency:

• Minimum radar cost

• Feasible timeline occupancy.

In examining the costs of phased-array radars, it is quickly discovered that larger wavelengths, that is, lower operating frequencies, exhibit much lower radar costs than their higher-frequency counterparts. This is due to two primary factors, including:

• Large effective antenna element area

• Higher power of transmit-receive (T/R) modules versus frequency.

The first factor results in fewer required antenna elements and T/R modules for a given size antenna aperture, and the second results in a higher transmit power for a fixed size antenna aperture. Together, these two factors strongly influence the radar front-end cost, that is, primarily the antenna and beam steering generator, which can represent one third to one half of the phased-array radar's total cost.

The search timeline occupancy consideration also favors lower operating frequencies, since for a given size antenna aperture, lower frequencies result in larger antenna beamwidths. This, in turn, results in many fewer beam positions required to search a given solid angle, the parameter Ψ in equation (10.1). This reduction in the number of beam positions (or "beams") corresponds to a lower beam rate to execute a given sized volume search, which can dramatically relax waveform scheduling requirements. Ultimately, the use of lower operating frequencies for search can reduce the timeline occupancy limitation suffered at higher frequencies. This is one reason for many search radars using the UHF or L-band radio frequency (RF) bands. Figure 10.1 illustrates the required search beam rate for different operating frequencies, where D is the antenna aperture diameter.

As can be seen from Figure 10.1, for a given antenna aperture diameter, the required search beam rate increases for higher operating frequencies (i.e., smaller wavelengths). Table 10.1 illustrates the trade-off for a volume search example with coverage:

• Range: 400 km to 1100 km

• Azimuth: 45 degrees

• Elevation: 25 degrees

• Search frame time: 0.5 seconds.

The waveform selected is a 2.5 ms pulse with a scheduling interval of 10 ms (i.e., 25% duty factor). The available beam rate for this waveform is 100 beams/second, and the antenna diameter is 5 meters. Table 10.1 indicates the advantage of lower operating frequency for timeline occupancy in this example.

Table 10.1 Search Timeline Occupancy vs. Operating Frequency Trade-Off

Operating Frequency	Required Beam Rate	Available Beam Rate	Occupancy Limited
UHF	25	100	No
L-band	136	100	Yes
S-band	273	100	Yes

10.2.2 Horizon Fence Search

Horizon fence searches can be used by radars to detect targets that "break the horizon." They are energy efficient searches since they only survey an azimuth sector one (or a few) beams high in elevation, usually at or slightly above (typically, 1 to 3 degrees) the local horizon. The RRE for this type of search is also provided in Chapter 1 and is defined as:

$$SNR = \frac{\sigma}{\left(2\sqrt{\pi}\right) k T_s \, R^3 \, \Psi \, N \, v_T \, L_t \, L_r} \frac{P_{AVE} \, A_r}{\sqrt{G_r}},$$ (10.2)

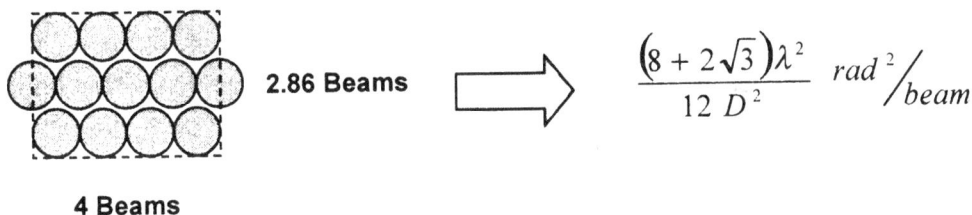

2.86 Beams ⟹ $\dfrac{\left(8 + 2\sqrt{3}\right)\lambda^2}{12 \, D^2} \; rad^2\!\Big/\!beam$

4 Beams

$$Required \; Beam \; Rate \; = \; \frac{\Psi}{T_{SC}} \frac{12 \, D^2}{\left(8 + 2\sqrt{3}\right)\lambda^2}$$

Figure 10.1 Required Search Beam Rate versus Operating Frequency

where N, v_T and G_r are the number of desired detection opportunities, target vertical velocity, and receive antenna gain, respectively. As can be seen, for horizon fence searches, there is a weak dependence on operating frequency via the antenna receive gain term, where:

$$G_r = \frac{4\pi A_r}{\lambda^2}. \tag{10.3}$$

Therefore, there is a small improvement in search sensitivity (i.e., SNR) for higher operating frequencies, which favors the selection of higher operating frequencies.

10.2.3   Tracking

To achieve tracking accuracy requirements, the appropriate form of the RRE is, again from Chapter 1, given by:

$$\sigma_\theta^2 = \frac{(4\pi)^3}{2 k_m^2 T_t} \frac{k T_s R^4 L_t L_r}{P_{AVE} A_r G_t G_r \sigma}. \tag{10.4}$$

As can be seen, accuracy improves as the inverse of the average power-aperture-gain squared product. Therefore, there is a very strong (f^4) dependence on frequency through the antenna gains, as defined in equation (10.3). If tracking is a primary requirement for the radar, then higher operating frequencies should be selected.

10.2.4   Target Classification and Discrimination

For these radar functions, high SNR is the driving requirement. In general, all target features used for classification and discrimination exhibit accuracies that improve with higher SNR.

The RRE that governs sensitivity (sometimes referred to as track sensitivity) is, as defined in Chapter 1, expressed as:

$$SNR = \frac{P_t G_t A_r \sigma}{\left(4\pi R^2\right)^2 k T_s B L_t L_r}. \tag{10.5}$$

As seen in equation (10.6), SNR is a function of the peak power-aperture-gain product. Similar to the tracking case, this relationship is frequency-dependent due to the antenna gain term defined in equation (10.3).

Therefore, sensitivity is proportional to the square of operating frequency, and improves at higher radar bands. Coupled with the f^4 relationship for track accuracy, which is fundamental to high-quality kinematics features, it can be seen why radars that require good target classification and discrimination performance will usually operate at higher operating frequencies.

10.2.5 Operating Environments

Typical phased-array radar operating environments, in addition to a clear environment, include:

- Surface and volumetric clutter

- Chaff

- Electronic counter measures (ECM) or jamming.

Each of the above environments will degrade the performance of all radar functions (e.g., search, track, target classification). As before, performance degradation can be predicted via the appropriate forms of the RRE.

10.2.5.1 Clutter. For surface or area clutter, the RRE is:

$$SCR = \frac{P_t G_t \sigma A_r}{\left(4\pi R^2\right)^2 L_t L_r} \frac{\left(4\pi R_c^2\right)^2 L_t L_r}{P_t G_t \sigma_{clutter} A_r^{'}}$$

$$= \left(\frac{A_r}{A_r^{'}}\right)^2 \left(\frac{R_c^3}{R^4}\right) \left(\frac{\sigma}{\sigma^{\circ}\left(\frac{c\,\tau}{2}\right)\tan\phi\ \theta_{AZ}}\right), \qquad (10.6)$$

where R_c, σ°, A'_r, τ, ϕ, and θ_{AZ} and are the range to the clutter, clutter coefficient, effective receive aperture to the clutter, pulse length, depression angle, and antenna 3 dB beamwidth in azimuth, respectively.

The RRE for volumetric clutter is:

$$SCR = \frac{P_t G_t \sigma A_r}{\left(4\pi R^2\right)^2 L_t L_r} \frac{\left(4\pi R_c^2\right)^2 L_t L_r}{P_t G_t \sigma_{clutter} A_r'}$$

$$= \left(\frac{A_r}{A_r'}\right)^2 \left(\frac{R_c^2}{R^4}\right) \left(\frac{\sigma}{\sigma^\circ \left(\dfrac{c\,\tau}{2}\right)\theta_{AZ}\,\theta_{EL}}\right), \qquad (10.7)$$

where θ_{EL} is the antenna 3 dB beamwidth in elevation.

Note in both equations (10.9) and (10.10) that frequency does not appear explicitly as a parameter. However, the antenna beamwidths decrease at higher frequencies and the clutter coefficient, σ°, is in general frequency dependent. Also, note that due to the Doppler relationship:

$$f_D = -\frac{2\dot{R}}{\lambda} \qquad (10.8)$$

at higher operating frequencies, the Doppler spread of the clutter is wider and, thus, higher pulse repetition frequencies (PRFs) will be required to maintain a reasonable clear Doppler region for target detection purposes. Selection of the PRF will have an impact on the range ambiguity of target detections (i.e., how many possible range intervals will have to be considered) and clutter fold-over effects. Therefore, operating frequency should be selected to balance these factors.

10.2.5.2 Chaff. Chaff, or large quantities of reflective materials (i.e., dipole antennas), is dispersed by a target to present an alternate large RCS target in the attempt to decoy the radar. It can be considered to be like volumetric clutter. Here, a mitigation technique for the radar can be to achieve the smallest resolution cell possible to limit the number of chaff dipoles that compete with the target RCS, or to let the difference in target and chaff range-rates separate the two. However, exploiting the difference in Doppler between the desired (i.e., the target) and undesired return (i.e., the chaff) can often prove to be adequate mitigation.

The radar has four fundamental resolution dimensions available for mitigating chaff:

• Range

• Azimuth and elevation angles

• Doppler.

Due to the size of the antenna beamwidth, which determines the radar's angular resolution, azimuth and elevation resolution are less useful for limiting the chaff volume. However, range and Doppler resolution can be more effective in limiting the chaff RCS that competes with the target before the target and chaff separate.

Since range resolution is a function of the waveform bandwidth, and, in general, higher operating frequencies allow wider bandwidths, operating at higher frequencies can enable better mitigation of chaff. In the Doppler dimension, resolution is driven by coherent integration time or the coherent processing interval (CPI):

$$\delta_{Doppler} = \frac{1}{T} = \frac{1}{CPI}. \tag{10.9}$$

The maximum CPI depends on the coherency time of the radar hardware, effectively established by the exciter, and the coherency time or correlation time constant of the target. In most cases, the target is the limiting factor. Enhancing Doppler resolution allows more rapid separation of targets and chaff.

10.2.5.3 Electronic Counter Measures. From Chapter 1, the RRE for the barrage noise jamming case defines the signal-to-interference ratio (SIR) to be:

$$SIR = \frac{P_t\, G_t\, G_r\, \lambda^2\, \sigma\, R_J^2}{\left(4\pi\right)^2 P_J\, G_J\, R^4\, L_t\, L_r}, \tag{10.10}$$

where P_J, G_J, and R_J are the jammer power, antenna gain, and slant range, respectively. Like the track sensitivity form of the RRE, this relationship can be re-expressed to be proportional to f^2, and therefore favors higher operating frequencies.

Hence, whether the jamming is in the radar's antenna mainlobe (i.e., a self-screening or escort jammer) or in an antenna sidelobe (i.e., a stand-off jammer), radars at higher operating frequencies achieve higher SIRs than lower frequency radars do when all other radar parameters have identical values.

10.2.6 Radar Applications

10.2.6.1 Air Defense. Air defense radars perform search and acquisition, tracking, and non-cooperative target recognition (NCTR). As shown in previous sections, the latter functions favor radars at higher operating frequencies.

10.2.6.2 Missile Defense. Radars that perform missile defense fire control must execute similar functions described for air defense radars.

10.2.6.3 Early Warning. Surveillance and target acquisition are the primary functions performed by early warning radars. As described in Sections 2.1 and 2.2, lower operating frequencies such as UHF and L-band are preferred frequencies for these functions.

10.2.6.4 Surface Target Search and Track. Typical surface search and track radars must operate in severe clutter environments. In order to mitigate clutter backscatter, either moving target indicator (MTI) or pulse-Doppler waveforms are employed. Since targets must be discriminated from stationary or slowly moving clutter, usually low to medium PRF waveforms are employed. Depending on additional functions to be performed by these radars, S-, C-, or X-band radars can be employed.

10.3 WAVEFORM SELECTION

A number of factors influence the selection of radar waveforms. The subsequent sections address a number of these factors.

10.3.1 Clear Environments

10.3.1.1 Search. In clear environments, either single-pulse or multiple-pulse waveforms can be used. Typically, narrow bandwidth waveforms are employed since search is a situational assessment function and is, therefore, only a "bell ringer" capability. Bandwidths in the range of several hundred kilohertz to around 1 MHz are used for search. Linear frequency modulation (LFM) is typically used. However, other modulations are possible, such as non-linear frequency modulation (NLFM) and phase-codes. One of the good qualities of LFM is that it is Doppler-tolerant; that is, there is small degradation at the output of the matched filter when knowledge of target-induced Doppler is uncertain (usually the case for search). For cases where a single pulse is inadequate for detection purposes, multiple pulses can be used, either to implement some form of non-coherent or coherent integration. Included in the former category are binary integration techniques such as M-out-of-N detection.

10.3.1.2 Track. Once targets are acquired, the track function can be initiated and maintained or updated. Depending on the type of targets to be tracked and the subsequent use of track data, different bandwidth waveforms can be employed. LFM pulses are typically employed, although as for search other modulations can be used such as NLFM and phase-coded waveforms. Again, like search, single or multiple-pulse waveforms are candidates for use. The latter type of waveform is used when single pulses do not afford adequate SNR. However, since tracking is based on likely detection at each update time, as opposed to cumulative detection probability as in the case of search, coherent integration waveforms are usually employed for tracking purposes in cases of low per-pulse SNR.

Heuristically, the track waveform bandwidth usually is selected to match the approximate length of the targets to be tracked. For air target tracking, bandwidths in the range 5 MHz to 20 MHz can be used. These bandwidths correspond to range resolutions of 30 meters to 7.5 meters, which encompass most air targets. For ballistic missile targets, and complexes of missile objects, wider bandwidths are typically employed in the range of 10 MHz to 50 MHz. These bandwidths correspond to range resolutions of 15 meters to 3 meters, which better match missile-type targets.

10.3.1.3 Target Classification and Discrimination. The waveforms required for target classification and discrimination are driven by the type of features extracted. In general there are two waveform types needed to collect the features used for target classification of air and ballistic missile targets: narrowband and wide-band. The narrowband waveforms are used for kinematics-based features (e.g., deceleration), as well as for signature features.

The bandwidths used for narrowband waveforms are similar to (and can be the same as) those used for tracking, for example, in the range of 5 MHz to 50 MHz. Wide bandwidths can be in the range of several hundred MHz and higher.

10.3.2 Clutter Environments

For situations where surface detection and tracking over land or sea is required, or for surface-based radars that must handle surface or low-elevation targets, severe clutter environments can exist. In these cases, the single-pulse waveforms discussed in Section 3.1 are not adequate since they do not possess clutter-mitigation capabilities. Instead, coherently processed, multiple-pulse waveforms are required. The common waveforms used for clutter environments are the MTI and pulse-Doppler types discussed in Chapters 1 and 2.

These are multiple-pulse coherent waveforms that generally consist of a sequence of three or more pulses, equally spaced in time (by the pulse repetition interval [PRI], or at a frequency referred to as the pulse repetition frequency or PRF). The term *coherent* refers to knowledge of phase of each pulse in the pulse train. Coherence is a required attribute of the waveform excitation or generation function of the radar. This function is usually performed in hardware by the exciter or transmitter.

MTI-type waveforms may consist of three to five pulses depending on the application, whereas pulse-Doppler waveforms can often consist of eight to thirty-two pulses. They are processed differently as described in Chapters 1 and 2.

10.3.2.1 Search. As for the clear environment case, search uses narrowband waveforms, typically in the range of several hundred kHz to 1 MHz or so. Each sub-pulse of the pulse train is often of the LFM type. The PRFs used are dependent on the expected target speeds and the Doppler extents of the clutter. However, it is desirable to use, if possible, PRFs low enough to be range unambiguous over the desired search range extent. This simplifies the processing necessary to extract target range. If higher PRFs are dictated to enable adequate clutter cancellation, then these range ambiguities will need to be dealt with to determine unambiguous target range.

10.3.2.2 Track. Tracking in clutter environments may use narrowband LFM sub-pulses as employed in the clear environment case with bandwidths in the range of 5 MHz to 50 MHz. In general, although low PRF waveforms with unambiguous range would be ideal, medium PRFs with some reasonable number of range ambiguities are acceptable. Often multiple, staggered PRFs are used to avoid blind ranges and speeds for both MTI and pulse-Doppler type processing.

10.3.2.3 Target Classification and Discrimination. Similar to the waveforms for tracking described above, medium PRFs with sub-pulse LFM bandwidths in the range of several hundred MHz and higher are employed for classification and discrimination purposes. In cases where Doppler is to be extracted, long coherent processing intervals can be used to improve Doppler resolution.

10.4 RADAR COVERAGE

Trade studies are often performed to optimize radar coverage in range, angle, and Doppler. Each type of trade-off analysis is described in the subsequent sections.

10.4.1 Range

Range coverage is often explicitly specified in radar requirements specifications. However, in some applications, only mission-level requirements are defined in these documents and radar coverage must be derived via trade studies and analysis.

Specific range coverage requirements can be a function of the type of mission (e.g., early warning, air defense, missile defense, data collection) or radar function (e.g., search, track, target classification or NCTR).

For air and ballistic missile early warning missions, the range coverage is usually interpreted as the search range extent as a function of azimuth, elevation, and altitude. For air and missile defense applications, necessary range coverage is derived to establish a defended area or keep-out zone, often for specific design scenarios. In many cases, maximizing battlespace is the driving requirement. In data collection applications, either explicit range coverage can be defined or data gathering scenarios are provided from which range coverage can be derived.

Surveillance functions usually have explicit range coverage requirements for autonomous capabilities such as volume searches or horizon fence searches. Sometimes these requirements can be derived from desired sensor-to-sensor handover or cued search capabilities. Similarly, when not explicitly specified, range coverage for tracking is driven by types and distributions of targets.

10.4.2   Angle

Azimuth and elevation coverage, when not explicitly specified in a radar system specification, can often be defined as described for range coverage in Section 4.1. Either mission or radar functions can establish these coverage requirements.

10.4.3   Doppler

Doppler coverage requirements can be established via trade studies to optimize the range of target velocities that the radar can handle in search and track. This coverage is a function of the radar operating frequency.

10.5   RECEIVER OPERATING CHARACTERISTIC DESIGN

The trade-offs required to establish receiver operating characteristic (ROC) parameters for phased-array radars are discussed in this section. In general, an ROC presents the probability of detection versus the probability of false alarm for different target, waveform, and processing cases of interest, usually depicted in a graphical format. Aspects of radar ROC design considered include:

• Target fluctuation types

• Probabilities of false alarm and detection

• Coherent and non-coherent integration.

10.5.1   Target Fluctuation Types

As described in Chapters 1 and 2, many different target RCS fluctuation models exist, such as:

• Constant RCS (i.e., nonfluctuating target)

• Chi-square models (e.g., Swerling models I through IV)

• Log-normal.

The first can represent isolated RF scatterers, whereas the second represents more complex target types (e.g., aircraft).

An example of an ROC that is defined analytically is the Swerling I RCS fluctuation model. The probability of detection is given by:

$$P_d = P_{fa}^{\frac{1}{1+SNR}},$$ (10.11)

where P_{fa} and SNR are the probability of false alarm and signal-to-noise ratio (SNR), respectively.

ROCs for the other target fluctuation types can be similarly represented, if not analytically as in equation (10.16), then by performing numerical integration of the appropriate target model's probability density functions. ROCs for common RCS fluctuation models are available in many radar detection theory textbooks [10, 12].

10.5.2   Probabilities of False Alarm and Detection

The trade-offs here typically involve selection of the probability of false alarm based on waveform bandwidth and processed range window to limit the number of false alarms per second reported to the radar's acquisition function. This number is in turn a function of the percentage of radar resources (e.g., duty factor, timeline occupancy) allocated to servicing false alarms. These resources refer to radar energy and timeline usage wasted by attempting to corroborate (or verify) and acquire (or initiate track of) detections due to noise detections. The trade-off is defined by the relationship:

$$P_{fa} = \frac{N_{FA} \, T}{N_b \, N_r},$$ (10.12)

where N_{FA}, T, N_b, and N_r are the false alarm rate, revisit time, number of antenna beams, and number of range cells, respectively.

As can be seen, the probability of false alarm is selected to limit false alarm rate (i.e., rate at which detection verification will occur), for the specified revisit time to a given antenna beam position, the size of the angular area covered, and the size of the range window (via the number of range cells and the waveform bandwidth).

10.5.3   Coherent and Non-Coherent Integration

The ROCs for single-pulse waveforms were discussed above. However, when pulse integration is considered to improve SNR for weak targets at long range, this affects the ROC. For example, when coherent integration is used, the resulting SNR is given by:

$$SNR_{CI} = \alpha \, N_{CI} \, SNR_1 ,$$ (10.13)

where α, N_{CI}, and SNR_1 are the coherent integration efficiency (less than unity), the number of target returns coherently integrated, and the single-pulse SNR, respectively. It should be noted that all pulses in the coherent pulse train must be at the same operating frequency.

For example, for the Swerling I illustration in Section 5.1, the ROC for coherently integrated returns is given by:

$$P_d = P_{fa}^{\frac{1}{1 + \alpha \, N_{CI} \, SNR_1}} .$$ (10.14)

In the case of non-coherent integration, a slightly different form of ROC is derived. Here it is advantageous to change frequency from pulse to pulse to decorrelate sequential target returns when targets follow a Swerling I or Swerling III RCS fluctuation model. This decorrelation transforms a Swerling I model into a Swerling II model, or alternatively, a Swerling III into Swerling IV. In either case, non-coherent integration of Swerling II or IV yield improved detectability as compared with Swerling I or III when per-pulse SNRs are greater than a minimum value.

10.6   SEARCH DESIGN

There are several possible trade studies to be performed to optimize the design of searches for phased-array radars, including:

• Target types, fluctuation models, and dynamics

• Search fence versus volumetric search

• Coherent and non-coherent integration

• Cumulative probability approach (e.g., binary, M-of-N).

These topics are covered in the following sections.

10.6.1   Target Types, Fluctuation Models, and Dynamics

As discussed in Chapters 1 and 2, there exists a number of different target RCS fluctuation models that attempt to represent the most common target types. These targets include:

• Air targets (e.g., aircraft, unmanned aerial vehicles [UAVs], cruise missiles)

• Ballistic missile targets

• Surface targets (e.g., vehicles, ships).

Each class of target exhibits different characteristics, including dynamics and RCS statistics. Air and surface target types are further described below.

10.6.1.1 Air Targets. Air targets are complex scatterers of RF due to their physical structure, shapes, and discontinuities (e.g., seams, appendages, sharp edges). These characteristics result in fluctuating RCSs as a function of viewing geometry, radar operating frequency, and waveform bandwidth. At bandwidths where multiple scattering sources are included in a single range resolution cell, constructive and destructive interference occurs due to the RF sources adding or subtracting as a result of the random-like phases (arising from target motion relative to the radar) associated with each unresolved scattering source. When bandwidths are large enough that only single scatterers are included in a range cell, no RCS fluctuations occur as long as there is no blockage or obscuration.

10.6.1.2 Surface Targets. Surface targets are very similar to air targets, except for their dynamics. Most surface targets such as ships and vehicles move at slow speeds, and so RCS fluctuations may be lower in frequency than for air targets. Again, simpler targets may fluctuate less rapidly (e.g., Swerling I or III) and more complex targets may fluctuate more rapidly (e.g., Swerling II or IV). Surface targets can possess large mean RCS values (e.g., 0 to 20 dBsm).

10.6.2 Search Fence versus Volumetric Search

Performing the surveillance or search capability is a key function for most radars since before any situation assessment, tracking, target classification, data collection, or other mission functions can begin, targets must be detected. Depending on the type of mission, the specific types of targets, and the size of the region to be surveyed, different kinds of search can be employed. A major trade-off is determining the most resource-efficient search technique to use for a given radar application.

The decision as to whether to use a horizon search fence versus a more comprehensive and resource intensive volume search is addressed here. The first basic question to be answered is whether a horizon fence is feasible from a mission perspective. For many ballistic missile early warning or fire control applications, a horizon fence is feasible.

However, for many air defense applications where large volumes need to be surveyed over an altitude regime that can include many diverse target types, a volume search might be the only practical solution. This may be the case for situation assessment purposes. Another less resource-intensive solution might instead be a number of stacked search fences, spaced appropriately in elevation to provide the necessary coverage without resorting to a complete volumetric search.

In any case, these and other alternatives need to be explored to select a search design where performance and radar cost are balanced. Coupled with this analysis are the selection of the specific detection technique for the target models to be encountered and any coherent or non-coherent pulse integration.

10.6.3 Coherent and Non-Coherent Integration

When additional detection capability is required for certain targets or applications, coherent or non-coherent integration can be useful. Examples of its use include acquiring very small targets at long ranges or performing other functions which require higher SNR than either the search or tracking functions. Integration is not normally used to overcome shortcomings in the radar sizing, but to allocate available radar duty factor and timeline occupancy for more stressing functions when necessary. This is preferable to driving the radar average-power aperture for a worst-case requirement when extra SNR is needed only occasionally or for certain modes of operation.

Since search performance is achieved via cumulative probability of detection, binary integration of individual pulses or coherently or non-coherently integrated batches of pulses are typically used. In general, non-coherent integration is used for search since for fluctuating targets this results in better detectability than attempting coherent integration on an inherently fluctuating RCS (i.e., since coherent integration is limited by the target RCS correlation time constant). Also, the phase coherence required for coherent integration is more difficult to insure.

The basic trade-off is whether for the type of targets being acquired whether frequency diversity should be employed to enhance detectability (e.g., to transform a Swerling I or III RCS model to a Swerling II or IV model, respectively). This can impose additional requirements on the search function, for example, by requiring that multiple frequencies be available for search, which might preclude search capability when only a few narrowband frequencies are available (e.g., due to interference).

10.6.4 Cumulative Probability Approach

Another trade-off topic is the type of detection processing and rules to use for search. Possibilities include:

• Single-pulse detection
• Binary integration (i.e., M-out-of-N rule)

- Non-coherent integration
- Coherent integration.

10.6.4.1 Single-Pulse Detection. The simplest approach is to use a single-pulse detection rule. In general, this is not advised as it leads to overdesigned radars. This occurs since the SNR required for single-pulse detection is much higher than that required when using multiple looks and a cumulative probability of detection approach. The higher SNR requirement translates to a pessimistic average power-aperture requirement, that is, an oversized radar can result.

10.6.4.2 Binary Integration. This is a common approach for declaring search returns. The requirement for detection is that M-out-of-N detection opportunities (i.e., "looks" at the target) are successful, that is, exceed the detection threshold. As covered in previous chapters, this results in a requirement for a lower per-pulse probability of detection, and therefore a lower per-pulse SNR to achieve the desired cumulative probability of detection. A special case of binary integration is a 1-out-of-N rule. This leads to the lowest required SNR but the highest probability of false alarm for a specified probability of detection.

10.6.4.3 Non-Coherent Integration. When per-pulse SNR requirements cannot be satisfied for either single-pulse or cumulative probability of detection approaches described in Sections 6.4.1 and 6.4.2, some form of pulse integration can be used. As previously discussed, non-coherent integration is often the best approach. NCI can be used with binary integration to efficiently achieve search requirements. This approach performs NCI of M batches of returns, and then applies the M-out-of-N rule. Alternatively, a single batch of returns could be used for detection. However, as discussed in Section 6.4.1, this can lead to an oversized radar.

10.6.4.4 Coherent Integration. This is similar to the approach in Section 6.4.3, except that coherent integration is substituted for NCI. As discussed above, this poses requirements on the coherency of the returns, which is again limited by the target's correlation time constant due to the expected fluctuation of the narrowband RCS for many types of targets.

10.7 TRACKING ARCHITECTURE AND PARAMETER SELECTION

Tracking trade-offs typically are performed for:

• Data association algorithm selection

• Tracking filter and target model selection,

as a function of target types and dynamics, and waveform parameters. These topics are discussed in the following sections.

10.7.1 Data Association Algorithm

Selection of an appropriate data association (DA) algorithm is a crucial decision in the design of a phased-array radar tracking function. Two factors that strongly influence this decision are the target characteristics (e.g., spatial density and dynamics) and the tracking waveform bandwidth (i.e., the range resolution of the waveform). These are not independent considerations. For typical narrow-band tracking, where the waveform bandwidth does not overresolve the target (i.e., isolate individual RF scatterers on the body of the target, or in other words results in a "point target"), the target density will be a function of the likely physical spacing of target objects. For example, air targets might not be densely spaced relative to the tracking waveform bandwidth.

The fundamental trade-off is the complexity of the selected DA algorithm versus the target density and expected target dynamics. In general, for air and surface target tracking simpler DA algorithms can usually be employed (e.g., nearest-neighbor [NN]). Alternatively, for closely spaced targets, more complex DA algorithms may be required (e.g., joint probabilistic data association [JPDA] or NN-JPDA, as discussed in other chapters of this book).

10.7.2 Tracking Filter and Target Model

In a similar manner, the type and dynamics of the targets to be tracked ultimately drive the selection of the class of tracking filter. Two major types of tracking are air and ballistic missile targets, with a third being surface targets including ships and vehicles. This section focuses on the first two since the third is less challenging and can be considered a subset of the air targets. Moreover, due to the relatively low-level dynamics of surface targets, for example, speeds and maneuvers, track-while-scan methods are often used, as are used for air-traffic control (ATC), which is covered in Chapter 1. The major issue for surface targets is mitigation of residual clutter returns.

10.7.2.1 Air Targets. Tracking many air targets, for example, aircraft and helicopters, has its challenges since all are either directly or remotely controlled by a pilot or operator. Therefore, unpredictable behavior, such as sudden maneuvers like dives, climbs, and high-g turns, must be accommodated by the tracking algorithms.

There are two basic approaches to tracking this class of target:

- Added process noise to accommodate unmodeled maneuvers
- Multiple-model tracking filters.

The classic use of the first approach is described in the Singer paper, reference [13]. Singer used a simple stochastic model to approximate the behavior of aircraft maneuvers. This model effectively modulates the level of process noise added to the system model to account for unmodeled aircraft turns, dives, and so on. If moderately accurate tracking performance is required, this approach works quite well. In addition, if better track accuracy is required and surplus radar resources are available, then increasing the track update rate and commensurately decreasing process noise will produce smaller tracking errors.

The second approach is to implement some form of multiple-model tracking filter, where several different models are employed, one for each specific class or magnitude of maneuver. A popular approach is the interacting multiple-model (IMM) discussed elsewhere in this book and in detail in [14–16]. The IMM tracking filter near-optimally blends the outputs of the different modeled filters. This approach can theoretically attain higher accuracy than adding process noise alone, and do so at lower tracking rates. In practice, the added-process noise approach usually satisfies air target tracking requirements.

10.7.2.2 Ballistic Missile Targets. In a similar fashion, there are also two fundamental approaches to tracking ballistic missile objects through all phases of flight: adding process noise, and modeling of multiple-phase dynamics using a tracking filter like the IMM technique. Again, aside from trading off increased track update rate for decreased process noise, for very accurate tracking, approaches such as IMM filters can be used. As for air targets, in practice, adding process noise is adequate for many applications.

10.8 TARGET CLASSIFICATION

The trade-offs performed in selection of appropriate target features and a classifier for phased-array radars are driven by the radar's mission or missions, and in turn by the types of targets to be classified. As for tracking, the three basic target classes are:

• Air targets

• Ballistic missile targets

• Ship and vehicle targets.

As discussed in other parts of this book, the Bayes' classifier is the optimal algorithm for deciding upon target classes or types. However, for many nonstressing applications, simpler methods such as decision trees can be employed when the targets classes are statistically well-separated. This is often the case when deciding among the three above target classes.

However, when deciding within target classes where there may be only small interclass statistical differences, algorithms such as the Bayes' or Dempster-Shafer approaches can be necessary. [17–21] are excellent sources of information on different classifiers and their applications.

In terms of trading off the value of competing target features, to a first order a K-factor analysis can be useful in determining which features have the most discriminating or separating power. Often only a small subset of the possible target features will be required to separate certain target classes or types. However, it is also true that the constituents of the subset will likely be different for separating different classes or types.

After preliminary screening of target features is performed using analytical means such as K-factor analysis, higher fidelity Monte Carlo analysis is generally necessary to finalize feature set composition and to refine classifier data-bases, and so on. These same Monte Carlo methods can then be used with different trials and randomly varied scenario and target parameters to predict classifier performance.

10.9 REFERENCES

This section provides a list of useful radar references for this chapter.

[1] J. V. Candy, *Signal Processing—The Modern Approach,* McGraw-Hill,

[2] S. Haykin & A. Steinhardt, *Adaptive Radar Detection and Estimation,* Wiley, 1992

[3] S. Haykin, *Adaptive Radar Signal Processing,* Wiley-Interscience, 2006

[4] S. Kay, *Modern Spectral Estimation: Theory and Application,* Prentice-Hall, 1999

[5] D. Manolakis, *Statistical and Adaptive Signal Processing,* Artech House, 2005

[6] S. L. Marple, *Digital Spectral Analysis with Applications,* Prentice-Hall, 1987

[7] R. A. Monzingo & T. M. Miller, *Introduction to Adaptive Arrays,* SciTech, 2003

[8] R. Nitzberg, *Radar Signal Processing and Adaptive Systems,* 2nd Edition, Artech House, 1999

[9] A. Oppenheim & R. Shafer, *Digital Signal Processing,* Prentice-Hall, 1975

[10] A. Papoulis, *Probability, Random Variables, and Stochastic Processes,* McGraw-Hill, 1965

[11] A. Papoulis, *Signal Analysis,* McGraw-Hill, 1977

[12] H. Van Trees, *Detection, Estimation and Modulation Theory, Part 1,* Wiley-Interscience, 2001

[13] R. A. Singer, "Estimating Optimal Tracking Filter Performance for Manned Targets," *IEEE AES-6,* Issue 4, July 1970, pp. 473–483

[14] Y. Bar-Shalom, *Multitarget-Multisensor Tracking: Principles and Techniques,* YBS, 1995

[15] Y. Bar-Shalom, *Multitarget/Multisensor Tracking: Applications and Advances,* Artech House, 2000

[16] S. Blackman & R. Popoli, *Design and Analysis of Modern Tracking Systems,* Artech House, 1999

[17] R. Duda, et al., *Pattern Classification,* 2nd Edition, Wiley-Interscience, 2000

[18] K. Fukunaga, *Introduction to Statistical Pattern Recognition,* 2nd Edition, Academic Press, 1990

[19] S. Theodoridis & K. Koutroumbas, *Pattern Recognition,* 2nd Edition, Academic Press, 2003

[20] P. Dempster, et al., *Classic Works on the Dempster-Shafer Theory of Belief Functions,* Springer, 2007

[21] G. Shafer, *A Mathematical Theory of Evidence,* Princeton University Press, 1976

11

Performance-Driven Radar Requirements

11.1 INTRODUCTION

A fundamental step in the design of phased-array radars is to allocate system-level radar requirements to hardware and software subsystems. This chapter addresses this topic for the hardware and software subsystems and components thereof that constitute the majority of phased-array radar systems.

11.2 RADAR HARDWARE REQUIREMENTS

The following sections address the allocation of requirements to hardware subsystems and components.

11.2.1 Radar Range Equation-Driven Requirements

11.2.1.1 Transmitter Peak Power. Equation (11.1) represents a standard form of the radar range equation (RRE):

$$SNR = \frac{P_t G_t G_r \lambda^2 \sigma}{(4\pi)^3 k T_s B R^4 L_{total}},$$

(11.1)

where the parameters are the familiar peak transmitter power, transmit and receive antenna gains, wavelength, RCS, Boltzmann's constant, system noise temperature, slant range to the target, and total radar losses, respectively.

Radar sizing trade studies usually produce requirements on equation (11.1), which can be rearranged as:

$$\frac{P_t G_t G_r}{T_s} = \frac{\left(SNR_{req}\right)\left(4\pi\right)^3 k\,B\,R^4 L_{total}}{\lambda^2 \sigma} .$$ (11.2)

The required peak power can be determined as part of the quantity on the left side of equation (11.2). The allocation to individual parameters is driven by several factors.

First, either the average power-aperture product or peak power-aperture product can be specified to achieve mission performance requirements for search, track, or target classification, whichever is most stressing. The associated track accuracy requirements will influence aperture size to obtain a sufficiently narrow antenna beamwidth, given by:

$$\theta_3 = \frac{0.886\,\lambda}{L} \approx \frac{0.886\,\lambda}{\sqrt{A_e}} = 1.77\sqrt{\frac{\pi}{G}} ,$$ (11.3)

for a square antenna, and therefore the required antenna gain is:

$$G = \left(\frac{1.77}{\theta_3}\right)^2 \pi .$$ (11.4)

When the same aperture is used on transmit and receive, and ignoring any antenna taper applied on receive, equation (11.2) can be expressed as:

$$\frac{P_t}{T_s} = \frac{\left(SNR_{req}\right)\left(64\pi\right)k\,B\,R^4 L_{total}}{\lambda^2 \sigma}\left(\frac{\theta_3}{1.77}\right)^4 .$$ (11.5)

Finally, noting the relationship of system noise temperature and receive noise figure:

$$T_s = \frac{0.876\,T_{sky} - 254}{L_a} + T_{tr}\left(L_r - 1\right) + F_n\,290 .$$ (11.6)

where T_{sky}, L_a, T_{tr}, L_r, and F_n are the sky temperature, antenna loss, transmission line temperature and loss, and low-noise amplifier noise figure, respectively. P_t can be calculated as:

$$P_t = \frac{\left(SNR_{req}\right)\left(4\pi\right)^3 k\,B\,R^4\,L_{total}}{\lambda^2\,\sigma\,\pi}\left(\frac{\theta_3}{1.77}\right)^4 T_s. \tag{11.7}$$

11.2.1.2 Transmit and Receive Antenna Gains. From equation (11.4), the basic antenna gains can be derived. These must be adjusted for antenna losses and any sidelobe taper applied on receive.

11.2.1.3 Noise Figure. Noise figure is usually driven by the first-stage low-noise amplifier (LNA) in the front-end receiver chain. Therefore, the specific microwave technology and front-end losses dominate in establishing the noise figure. The noise figure can then be used to calculate the system noise temperature. For active phased-array radars that employ solid-state transmit/receive (T/R) modules at the antenna front-end, the noise figure is established by the LNA technology used (e.g., GaAs, GaN), and the particular T/R module architecture and its design parameters.

11.2.2 Environment-Driven Requirements

A desirable attribute of radar front-ends is that they operate linearly over the full range of expected signal levels. Since input signals can be composed of target echoes, thermal noise, interference, clutter, and so on, the linear dynamic range (DR) must accommodate the minimum and maximum signal levels:

$$DR_{min} = \frac{S_{max}}{S_{min}}. \tag{11.8}$$

The linear dynamic range is often expressed in dB as:

$$DR_{min}(dB) = 10\log_{10}\left[\left(\frac{S_{max}}{S_{min}}\right)^2\right] = 20\log_{10}\left(\frac{S_{max}}{S_{min}}\right). \tag{11.9}$$

The receiver and analog-to-digital converter (A/D) must accommodate:

- The minimum dynamic range

- The system noise level

- Any desired headroom or margin.

This requires that an N bit A/D satisfies:

$$2^N = \left(10^{\frac{DR_{min}(dB)}{20}}\right)\left(10^{\frac{Margin\,(dB)}{20}}\right)\left(System\ Noise\ (LSBs)\right) \tag{11.10}$$

where DR_{min} is the minimum dynamic range in dB, *Margin* is the desired head-room in dB, and *System Noise* is the level of system noise in A/D least significant bits (LSBs). The number of bits required in the A/D is then:

$$N = int\left\{\frac{ln\left(10^{\frac{DR_{min}(dB)}{20}}\right)}{ln\ 2} + \frac{ln\left(10^{\frac{Margin\,(dB)}{20}}\right)}{ln\ 2} + \frac{System\ Noise\ (LSBs)}{ln\ 2} + 1\right\} \tag{11.11}$$

Consider two examples. The first example is for a missile defense application where clutter is not an issue and the dynamic range for the missile objects is 50 dB with a headroom of 10 dB and noise set at three counts or LSBs. Using equation (11.11), the required number of bits is:

$$N = 12\ bits.$$

The second example is shipboard radar for and ship self-defense (SSD) mission. Here, the dynamic range is driven by surface clutter. If analysis shows that a 77 dB clutter-to-noise ratio (CNR) exists, a 15 dB margin or A/D headroom is to be allowed, and the receiver noise is set at three counts (i.e., three LSBs) on the A/D, then again using equation (11.11), the number of A/D bits can be calculated as:

$$N = 17\ bits.$$

11.2.3   Waveform-Driven Requirements

11.2.3.1 A/D Sampling Rate. The A/D sampling rate is established by the radar waveforms and processing. In traditional receivers that employ synchronous detection at baseband that results in complex-valued signals at the receiver output, that is, the A/D, the required sampling rate for the in-phase (I) and quadrature (Q) channels is set by the maximum baseband signal bandwidth.

The sampling rate is established by the radar waveforms and processing. Traditional receivers with synchronous detection with complex-valued signals at baseband, the I and Q sampling rate is:

$$f_{sampling} \geq k_{os} B_{baseband} \tag{11.12}$$

where k_{os} and $B_{baseband}$ are the over-sampling factor (relative to the minimum Nyquist rate, where $k_{os} \geq 1$ to allow range and amplitude interpolation), and the bandwidth of the baseband signal, respectively.

Therefore, the A/D sampling rate for non-stretch processed waveforms is:

$$f_{sampling} \geq k_{os} B_{mod} \tag{11.13}$$

where the baseband signal bandwidth is the waveform modulation bandwidth B_{mod}.

For the case of wideband waveforms and stretch processing, the sampling rate is:

$$f_{sampling} \geq k_{os} B_{dechirp} \tag{11.14}$$

where $B_{dechirp}$ is the bandwidth of the LFM waveform after the receiver's de-ramp down-converter.

Therefore, the A/D sampling rate for nonstretch processed waveforms is, from equations (11.13) and (11.14), given by:

$$f_{sampling} = k_{os} k_{IF} B_{mod} . \tag{11.15}$$

For the case of wideband waveforms and stretch processing, equation (11.14) becomes alternatively,

$$f_{sampling} = k_{os} k_{IF} B_{dechirp} ,$$
(11.16)

where $B_{dechirp}$ is the bandwidth of the LFM waveform after the receiver's de-ramp down-converter.

Examples of waveforms that are to be digitally pulse compressed and stretch processed are given in Table 11.1. The oversampling factor (k_{os}) is usually selected to allow range and amplitude interpolation after pulse compression (i.e., pulse matched filtering) or stretch processing.

Table 11.1   Example A/D Sampling Rate

	Digital pulse compression	Stretch processing
Over-sample factor, k_{os}	1.2	1.2
LFM or de-ramp bandwidth, B_{mod} or $B_{dechirp}$	10 MHz	50 MHz (1 ms @ 500 MHz; 15 km) (0.5 MHz/µs × 100 µs)
A/D sampling rate, $f_{sampling}$	12 MHz	60 MHz

11.2.4 Clutter Cancellation-Driven Requirements

11.2.4.1 Phase Noise. Phase noise requirements are usually established by the clutter cancellation requirements. In cases where MTI or pulse-Doppler waveforms and processing are used for clutter mitigation, the total integrated phase noise, primarily on the waveform exciter side (although depending on the transmitter type, the T/R module may also contribute), as well as in some cases the receiver, depending on the source of stable local oscillator (STALO) for the receiver and exciter, will limit possible clutter cancellation.

In most cases, the capability to cancel clutter is limited to:

Clutter Cancellation Ratio (dB) = Integrated Phase Noise Level (dBc) + M_{dB} ,

(11.17)

where the phase noise level is measured in dB relative to the carrier (dBc), and is the desired cancellation margin. Therefore, the maximum integrated phase noise level is given by:

$$Integrated\ Phase\ Noise\ Level\ (dBc)\ \leq\ Clutter\ Cancellation\ Ratio\ (dB)\ -\ M_{dB}\ .$$
(11.18)

For example, using equation (11.18) to calculate the integrated system phase noise level (i.e., from all contributions: e.g., exciter, receiver, antenna), for a required cancellation ratio (CR) of 50 dB with a desired 10 dB margin results in an integrated phase noise requirement of:

$$Integrated\ Phase\ Noise\ Level\ (dBc)\ \leq\ -50\,dBc - 10\,dBc = -60\,dBc\ .$$

This level, in turn, defines the allowable phase noise spectral content or frequency response (in dBc) versus distance from the carrier. This allowed spectral characteristic is suballocated to all contributing components, and is specified in the requirements documents for these elements (e.g., exciter, receiver, antenna).

11.2.5 Interference Cancellation-Driven Requirements

11.2.5.1 Amplitude and Phase Errors. These errors will limit the ability to accurately estimate and cancel interference, including both intentional (e.g., jamming) and unintentional (co-site or inter-radar, or radio interference sources). In some cases, these errors will also limit the performance of some target classification features.

The relationship between amplitude error level and maximum interference level is given by:

$$Amplitude\ Error\,(dB)\ =\ 20\,\log_{10}\left[\left(1+\delta/2\right)/\left(1-\delta/2\right)\right]$$
(11.19)

or, equivalently:

$$Amplitude\ Error\,(dB)\ =\ \left[20/\ln 10\right]\ln\left[\left(1+\delta/2\right)/\left(1-\delta/2\right)\right]$$
(11.20)

For small amplitude errors, the relationship for the error variance is approximately:

$$V_{\mathcal{E}} = E\{\varepsilon^2\} = \left[Amplitude\ Error\,(dB)/30\right]^2 \qquad (11.21)$$

As an example, for a desired CR of 40 dB, $V_{\mathcal{E}}$ = 0.0001, so the peak-to-peak variation across the filter band must be less than 0.3 dB. Table 11.2 provides the achievable cancellation versus amplitude errors for a sinusoidal amplitude ripple, assuming perfect receiver channel-to-channel matching (i.e., perfect alignment).

Table 11.2   Cancellation Ratio versus Peak Sinusoidal Amplitude Error

Peak Sinusoidal Amplitude Error (dB)	Achievable Cancellation Ratio (dB)
0.02	52.7
0.05	44.8
0.1	38.7
0.2	32.7
0.3	29.1

Allowable phase errors can be related to cancellation ratio by the approximate relationship:

$$Angle\ Error\,(\mathrm{deg}) = \left(\frac{180}{\pi}\right)10^{\frac{Error\,(dB)}{20}} \qquad (11.22)$$

11.2.5.2 Channel-to-Channel Alignment. A second limiting factor for achievable interference cancellation is the alignment and calibration of the receiver channels used by the particular cancellation approach, for example, sidelobe cancellers (SLCs), multiple SLCs, or adaptive arrays, as described in Chapter 8. The approximate relationship between channel-to-channel alignment errors and interference cancellation is given by:

$$CR = \left(\frac{4}{3}\right)\left(\varepsilon_p - 1\right)^2 \qquad (11.23)$$

where ε_p is the peak transfer function error. Table 11.3 illustrates available cancellation versus channel matching, that is, residual peak mismatch error, assuming no amplitude and phase errors as described in Section 11.2.6.1.

Table 11.3   Cancellation Ratio versus Peak Amplitude Mismatch Error

Peak Amplitude Mismatch Error (dB)	AchievableCancellation Ratio (dB)
0.02	51.5
0.05	43.5
0.1	37.5
0.2	31.4
0.3	27.8
0.4	25.3
0.5	23.3

11.2.6   Processing Throughput

Processing throughput requirements are driven by the complexity and amount of data that must be accommodated by the radar data processor. Here both signal processing and data processing algorithms are being considered. Hence the necessary computer throughput capacity required is a function of several parameters, including:

• Radar functions being performed

• Waveforms used

• Matched filtering employed

• Numbers of detections per unit time

• Number of tracked objects and update rates

• Types of interference cancellation algorithms.

These are covered in more detail in Section 11.3.

11.3   RADAR PROCESSING SOFTWARE REQUIREMENTS

11.3.1   Overview

The major radar processing software includes four categories:

• Mission software

• Signal processing software

• Diagnostics and test software

• Simulation software.

These are described briefly in the following sections.

11.3.1.1 Mission Software. For the most part, computer throughput required for the mission software is driven by the number and rate of waveforms to be scheduled, the complexity of the scheduling algorithms, the complexity of search processing, the number of concurrent tracks, the complexity of the data association and tracking algorithms, and the complexity of any target classification and discrimination algorithms needed for the particular mission.

11.3.1.2 Signal Processing Software. In a similar fashion, processing throughput required for signal processing is driven by the number of detections per unit time, the types of digital matched filtering employed (per receive action), constant false alarm rate (CFAR) processing (per range cell), Doppler processing (per Doppler filter), MTI processing (per range cell), detection processing (per range cell), and postdetection processing, such as peak detection (per range cell), interpolation (per detection), and monopulse processing (per track detection).

11.3.1.3 Diagnostics and Test Software. There are a few types of diagnostic and test software functions, including:

• Periodic health assessments (e.g., waveform tests such as for range sidelobes)

• Polling of hardware resident built-in-test (BIT) results and status

- Manual operation of the radar hardware

- Real-time calibration and alignment (e.g., pilot pulse measurements and compensation/correction processing).

Most of these are run in the background and by, at a low rate, stealing radar resource intervals for pilot pulse injections, waveform tests, and so on. This processing tends to be fairly low in required computer throughput.

11.3.1.4 Simulation Software. This function is usually used for initial integration and test of the mission software and for checking interfaces, and so on. The driver here is the number of real-time target injections to be performed as controlled by a scripted scenario. Whether this injection is performed digitally (i.e., directly into the mission software returns handler) or as RF test target injections into the receiver front-end, the basic processing is driven by the number of concurrent targets included in the predefined scenario, which is converted into a real-time sequence of radar return data.

11.3.2 Track-Driven Requirements

11.3.2.1 Track File Capacity. One of two situations usually establishes the track file capacity requirements:

- A direct requirement in the system specification

- Derived requirements to satisfy mission requirements, including analysis of typical or specified scenarios.

In the first case no requirements allocation is required. The second case necessitates analysis of mission requirements, expected targets densities and entry-rates, specific target scenarios, and so on, to derive the likely number of possible concurrent tracks. This analysis can be used as the basis for establishing a track file capacity requirement. The reality that some finite number of false tracks due to noise, clutter, or interference will exist, and also a number of redundant tracks, coasted tracks, and so on, invariably will arise, which together will necessitate augmenting the capacity derived from the above analysis by an adequate amount to accommodate them.

11.3.2.2 Track Update Rate. The track update rate is driven by the specified track accuracy requirement once the signal-to-noise ratio (SNR) and track time are established for a radar system with a fixed waveform bandwidth and antenna beamwidth. Obviously, the types of targets to be tracked will also impact the tracking accuracy.

Equation (11.28) is an approximate relationship for tracking accuracy:

$$\sigma_p^2 = \frac{R\,\theta_3}{2\,SNR\,f_r\,T}, \tag{11.24}$$

where R, θ_3, f_r, and T are the slant range of the target, 3 dB antenna beamwidth, number of track update rate, and time-in-track, respectively. Solving for the track update rate yields an approximate expression:

$$f_r = \frac{R\,\theta_3}{2\,SNR\,\sigma_p^2\,T}. \tag{11.25}$$

11.3.2.3 Data Association Capability. The driving requirement for data association algorithms is the real target density and the operating environment. The environment will establish what other target-like objects will be present along with the real targets such as residual clutter returns, chaff, deceptive jamming, and so on. The target types and the track waveform bandwidth will establish the number of RF scatterers per object and spacing of objects (or scatterers). Based on these target characteristics and environment, requirements on the type and quality of data association algorithms can be specified.

The possible target types will depend on the radar's mission. For example, for air targets with aircraft, UAVs, helicopters, cruise missiles, and so on, operating in a clear environment, except for rare circumstances, a relatively simple data association algorithm can be used. When clutter, chaff, or in other more stressing environments, the more complex data association algorithms will often be required (e.g., near-neighbor-joint probability data association [NN-JPDA], multiple hypothesis tracker [MHT], multidimensional assignment, Koch MHT).

11.3.3   Target Classification-Driven Requirements

11.3.3.1 Target Classification Capability. The requirements on target classification are driven for the most part by a few considerations, including:

- Classes and/or types of targets to be classified

- Radar operating frequency and waveform bandwidths

- Available target features

- SNR

- Required probabilities of correct classification and misclassification.

The target classes and/or types and their statistically defined differences are the major driver as to how difficult the classification problem is for given probabilities of correct classification and misclassification. The radar operating frequency, waveform bandwidth, and SNR are closely associated with the available target features and their quality, which influence the feasibility and performance in satisfying the probability of correct classification and misclassification.

In general, for target classes where there exist large statistical separations between classes, the classification or discrimination problem is simpler and less sophisticated classifiers can be employed, such as decision trees or straightforward logic (e.g., if-then-else constructs, or simple neural networks). For applications such as missile defense where target types may exhibit smaller statistical separations, more sophisticated approaches like Bayes' classifiers or Dempster-Shafer evidential reasoning may be required.

No simple analytical approaches exist to select the a near-optimal or appropriate classifier and target feature sets. Back-of-the-envelope analyses such as using K-factors with simple Gaussian feature statistics can be used for initial trades and preliminary classifier and feature selection. However, full-up Monte Carlo simulations with accurately-modeled waveform parameters, target characteristics, dynamic scenarios, feature extraction models, and so on, are ultimately necessary to derive reasonable and feasible classifier requirements.

11.3.3.2 Target Feature Extraction Capability. As described in Section 11.3.3.1, performance trade-offs are required to select the right classifier and target feature set. Selection of the target feature set will establish the requirements on feature extraction given the operating frequency of the radar systems and the waveform parameters.

11.3.4   Signal Processing-Driven Requirements

11.3.4.1 Signal Processing Throughput. Signal processing throughput, as described in Section 11.3.1.2, is largely driven by the number radar actions per second, the types of digital matched filtering employed (per radar receive action), CFAR processing (per range cell per radar receive action), Doppler processing (per Doppler filter per radar receive action), MTI processing (per range cell per radar receive action), detection processing (per range cell per radar receive action), and postdetection processing, such as peak detection (per range cell per radar receive action), interpolation (per detection per radar receive action), monopulse processing (per track detection per radar receive action), and so on.

Each of the above functions can be characterized using numbers and types of computations necessary for their implementation. By multiplying these by the relevant radar receive action rates, and summing over all major signal processing functions, an estimate of required throughput can be calculated.

11.3.4.2 Fast Fourier Transform (FFT) Sizes. Fast Fourier transform (FFT) size requirements are driven by the matched filter and Doppler processing employed, the A/D sampling rates, and sizes of range windows for each of the waveforms employed by the radar to execute its major functions (search, track, target classification and discrimination, etc.). An approximate FFT size for a given function is given by:

$$N_{FFT} = modulo\, power-of-two \left[\frac{2\, f_{sampling}\, RW}{c} \right], \qquad (11.26)$$

where $f_{sampling}$, RW, and c are the A/D sampling rate, receive window for given function's waveform, and the speed of light, respectively. The modulo function is defined to pick the power of 2 greater than or equal to its argument.

For example, if the argument in equation (11.30) is 10,525, the selected FFT size would be 16K, where the remaining 16,384 − 10,525 = 5,859 FFT inputs would generally be set to zero (i.e., referred to as zero filling).

11.3.4.3 CFAR Types. Selection of the appropriate CFAR type depends on the radar functions, target types, environment, waveform resolution, and other factors.

The detection logic, which includes the use of the CFAR processor, is defined in the following sections. In general, the logic used is dependent on the radar function: that is, search, track, or wideband data collection for target classification. In general, the same set of CFAR logic and parameters are used for most radar's detection processing for a given function.

Search. *Ballistic Missile Search.* In search, the targets are typically at the longest slant range from the radar. This results in low signal-to-noise ratios for the small radar cross section (RCS) targets that the radar must detect. For this reason, it is important to minimize any detection losses when performing long-range searches. Therefore, the rationale used for search is to use a constant noise threshold as the primary detection threshold, with the CFAR-generated threshold used as the secondary threshold. This approach will inherently minimize detection losses, since a loss will only be incurred when the CFAR threshold is used. Additionally, a linear CFAR algorithm is employed, since only a relatively small range of targets is expected, and a smaller loss is incurred using the linear CFAR than using nonlinear approaches such as logarithmic CFAR. Using linear CFAR is a reasonable approach when the expected dynamic range is small.

The logic employed selects the larger of the noise or a biased-CFAR threshold. The bias is selected to insure that the CFAR-generated value is only used when the noise threshold is overly optimistic. The logic can be expressed as:

$$\text{Threshold } T = \text{ Greatest of } \left\{ \text{Noise Threshold or } \beta \text{ Linear} - \text{CFAR Threshold} \right\} ,$$

$$(11.27)$$

where example CFAR parameters are included in Table 11.4.

As can be seen, a relatively small probability of false alarm is specified for search. This is intended to minimize wasting radar resources (i.e., duty factor and timeline occupancy) that might result if verify and track initiation waveforms are scheduled in response to false alarms. The noise threshold is set at the ideal thermal noise value associated with the desired probability of false alarm. Greatest-of processing is selected to minimize noise detections. A nominal three-cell CFAR gap is specified to mitigate multiple detections from the main target response or first time-sidelobes after matched filtering.

Table 11.4   Search Logic Parameters

Parameters Name	Parameter Value
Noise threshold	$-2 \ln (P_{FA})$
Nominal P_{FA} for noise and CFAR	10^{-6}
β	0.9
CFAR type	Linear
CFAR window select	Greatest of leading/lagging
CFAR gap	3 cells
Censoring select	Enabled

Cued Missile Search and Satellite Search.   These types of searches are more like track in that due to the availability of a target state vector (a cue for missile-type handover acquisition and orbital element set [OES] based predictions for satellites), only limited-range windows and track waveform bandwidths are employed. Therefore, linear CFARs can be employed. The parameters are essentially identical to the tracking cases described under "Track."

Track.   *Ballistic Missile Track.*   For tracking purposes, the CFAR-generated threshold is used exclusively. If a very large dynamic range of target RCS values are expected, then a logarithmic CFAR can be used. Due to the relationship between the logarithm of Rayleigh noise versus the sum-of-logs, a 2.5 dB bias is applied to the CFAR computed threshold. The detection logic for tracking with a logarithmic CFAR is given by:

$$Threshold \ T \ = \ \alpha \ (Log - CFAR \ Threshold) \ , \tag{11.28}$$

where example parameters are included in Table 11.5. In cases when a linear CFAR is preferred, the threshold is given by:

$$Threshold\ T\ =\ Linear - CFAR\ Threshold\ , \tag{11.29}$$

where example values are included in Table 11.6.

As can be seen for tracking, a relatively larger probability of false alarm is

Table 11.5 **Track Logic Parameters for Logarithmic CFAR**

Parameter Name	Parameter Value
α	$10^{(2.5/20)}$
P_{FA} for CFAR threshold	10^{-4}
CFAR type	Logarithmic
CFAR window select	Greatest of leading/lagging
CFAR gap	3 cells
Censoring select	Enabled

specified than for search. This is to maximize small target detectability since the

Table 11.6 **Track Logic Parameters for Linear CFAR**

Parameter Name	Parameter Value
P_{FA} for CFAR threshold	10^{-4}
CFAR type	Linear
CFAR window select	Greatest of leading/lagging
CFAR gap	3 cells
Censoring select	Enabled

target's range uncertainty, and therefore the effective false alarm rate is much lower for tracking as compared to search, which employs large range windows. Greatest-of processing is again selected to minimize noise detections. A nominal three-cell CFAR gap is specified to mitigate multiple detections from the main target response or first time-sidelobes after matched filtering.

The selection of linear versus logarithmic CFAR is dependent on the expected dynamic range of targets and environmental returns such as clutter or interference. For high-elevation tracking, which is common, the dynamic range is driven by the relative radar cross sections of missile objects. Since this is generally small, a linear CFAR should be selected. However, for low-elevation tracking where large clutter returns may compete with small targets, use of the logarithmic CFAR can be employed. In low-elevation or look-down radar applications, surface clutter backscatter levels drive the necessary dynamic range, as discussed in Section 11.2.2.

Target Classification.   For wideband data collection for target classification purposes, a larger probability of false alarm is again specified. This is to maximize small target detectability since the target's range uncertainty is very small, and therefore the effective false alarm rate is much lower for target classification data collection as compared to search or track, which employ larger range windows. Greatest-of processing is again selected to minimize noise detections. A nominal three-cell CFAR gap is specified to mitigate multiple detections from the main target response or first time-sidelobes after matched filtering.

Rationale for Selection of CFAR Type.   The basis for selecting the type of CFAR as a function of frequency is minimizing the expected loss in detectability incurred when using a CFAR compared with an ideal noise threshold given by:

$$T_{noise} = \sqrt{2\sigma_n^2} \, \ln P_{FA} \, , \qquad (11.30)$$

where $2\sigma_n^2$ and P_{FA} are the total front-end thermal noise power and probability of false alarm, respectively. All CFARs incur a loss in detectability due to the use of a limited number of noise samples to compute the average background noise level. This is referred to as a CFAR loss. In general, the noise power estimation error variance decreases as $1/N$ where N is the number of statistically independent samples used to compute the average. Therefore, use of large CFAR windows decreases the CFAR loss, while features such as least-of, greatest-of, and cell-censoring increase CFAR loss. In addition, use of logarithmic CFAR incurs a higher CFAR loss than linear CFAR. Table 11.7 illustrates how losses vary for different types of CFAR for 32 cells and a probability of false alarm of 10^{-6} [8].

Based on results such as those shown in the table, selection of the CFAR types described in Section 11.3.4.3 can be justified. Since preserving sensitivity during search activities is desirable when there is a requirement to detect small RCS tar-

Table 11.7   CFAR Losses for 32-Cell CFARs
with P_{FA} of 10^{-6}

CA-CFAR	0.97 dB
GO-CFAR	1.13 dB
OS-CFAR (75% rank)	1.45 dB
GO-OS-CFAR (75% rank)	1.66 dB
CA-CFAR, 1-cell censor	1.01 dB
CA-CFAR, 2-cell censor	1.06 dB

gets at long ranges, a noise threshold is the selected detection threshold since it in theory incurs no CFAR loss. In reality, since a finite number of noise samples are used to estimate the background noise power, there is an SNR penalty, although it is usually negligible since a large number of samples are typically used for its computation (e.g., on the order of 1,000 or more).

As can be seen in Table 11.11, the cell-averaging (CA), and CA with 1 or 2 cell-censoring types suffer the smallest losses for a given CFAR window size, ranging between 0.97 dB and 1.06 dB for the 32-cell processor. These losses are acceptable for tracking and classifying ballistic missiles where sensitivity is less of an issue. For the boosting-object track and target classification, the leading or lagging CFAR (identical to greatest-of CFAR [GO-CFAR] type) incurs only a slightly larger loss (1.13 dB for the 32-cell window case).

11.3.4.4 Post-Detection Processing Capability. The functions described as post-detection processing typically include:

• Peak detection

• Range and amplitude interpolation

• Monopulse processing.

These are for the most part independent of radar application, except for perhaps the required interpolation accuracies to limit signal processing losses. A simple quadratic interpolator is used in most radar applications. This generally will achieve in the range of approximately 10-to-1 enhancement of the reported range and amplitude values. However, this does require data from adjacent range cells, and therefore necessitates an oversampling factor in the range of 20 to 25% above the minimum Nyquist rate value.

11.4 REFERENCES

[1] J. V. Candy, *Signal Processing—The Modern Approach,* McGraw-Hill,
[2] S. Haykin & A. Steinhardt, *Adaptive Radar Detection and Estimation,* Wiley, 1992
[3] S. Haykin, *Adaptive Radar Signal Processing,* Wiley-Interscience, 2006
[4] S. Kay, *Modern Spectral Estimation: Theory and Application,* Prentice-Hall, 1999
[5] D. Manolakis, *Statistical and Adaptive Signal Processing,* Artech House, 2005
[6] S. L. Marple, *Digital Spectral Analysis with Applications,* Prentice-Hall, 1987
[7] R. A. Monzingo & T. M. Miller, *Introduction to Adaptive Arrays,* SciTech, 2003
[8] R. Nitzberg, *Radar Signal Processing and Adaptive Systems,* 2nd Edition, Artech House, 1999
[9] A. Oppenheim & R. Shafer, *Digital Signal Processing,* Prentice-Hall, 1975
[10] A. Papoulis, *Probability, Random Variables, and Stochastic Processes,* McGraw-Hill, 1965
[11] A. Papoulis, *Signal Analysis,* McGraw-Hill, 1977
[12] H. Van Trees, *Detection, Estimation and Modulation Theory, Part 1,* Wiley-Interscience, 2001
[13] Y. Bar-Shalom, *Multitarget–Multisensor Tracking: Principles and Techniques,* YBS, 1995
[14] Y. Bar-Shalom, *Multitarget/Multisensor Tracking: Applications and Advances,* Artech House, 2000
[15] S. Blackman & R. Popoli, *Design and Analysis of Modern Tracking Systems,* Artech House, 1999
[16] R. Duda, et al., *Pattern Classification,* 2nd Edition, Wiley-Interscience, 2000
[17] K. Fukunaga, *Introduction to Statistical Pattern Recognition,* 2nd Edition, Academic Press, 1990
[18] S. Theodoridis & K. Koutroumbas, *Pattern Recognition,* 2nd Edition, Academic Press, 2003

12

Missile Defense Radar Design Considerations

12.1 INTRODUCTION

This chapter treats some of the key aspects of missile defense radar design. Topics covered include:

- Missile defense mission parameters and requirements
- Ballistic missile threat types:
  - Missile systems
- Interceptor capabilities:
  - Maximum velocity
  - Fly-out range
  - Divert capability
  - Interceptor support requirements
- Desired defended area
- Radar requirements:
  - Search requirements

- Tracking requirements
- Target features
- Classifier requirements
- Waveform requirements

• Performance evaluation and design validation

Ballistic missile defense (BMD) radars are fire control radars where the intended target is a ballistic missile, such as a tactical ballistic missile (TBM), intermediate range ballistic missile (IRBM), or intercontinental ballistic missile (ICBM), and the defensive weapon is an interceptor missile. Figure 12.1 illustrates the CONOPS or mission of a shipboard BMD radar, also referred to as the DoDAF OV-1 architecture view for this type of system.

A BMD radar typically performs the search, acquisition, tracking, target classification and discrimination, and interceptor support functions. As such and due to the specific characteristics of ballistic missile targets, each of these functions is specialized for this type of threat.

For example, most BMD radars perform an autonomous horizon search fence to initially detect and acquire missiles, in addition to being cued by other

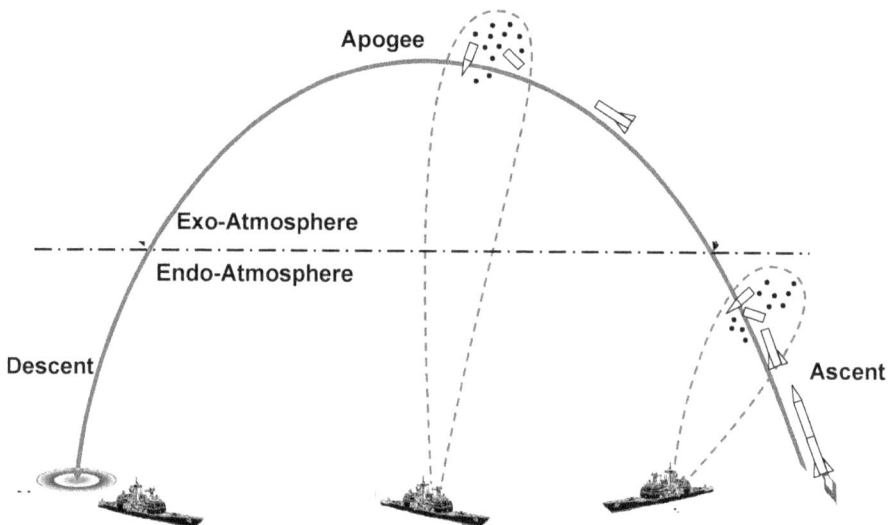

Figure 12.1   Mission of Shipboard BMD Radar (OV-1)

sensors (e.g., a handover from another radar or electro-optics sensor). A horizon search fence is a resource-effective form of search that is customized for acquiring ballistic missiles. Since it avoids performing a large brute-force volume search, the radar can be sized more modestly than, say, and anti-air warfare (AAW) or air defense (AD) radar, which must consider threats at varying ranges and altitudes over a large volume of space. Assuming that the BMD radar is appropriately sized for the threats it must operate against, no missile can over-fly a horizon fence. As shown in Chapter 1, there is a special form of the radar range equation for a horizon fence search that weakly favors lower operating frequencies.

Tracking of ballistic missiles differs from other types of targets in that long-range missiles have three distinct phases of flight:

- Ascent or boost phase (rocket booster is accelerating while within the earth's atmosphere)

- Midcourse or ballistic phase

- Descent or reentry (missile objects reenter earth's atmosphere and are subject to drag deceleration).

Tracking filters, therefore, must operate over one or more of these flight regimes, depending on the role of the particular BMD system.

Target classification and discrimination of ballistic missiles during the above phases of flight uses different approaches, depending on whether objects are powered (in boost phase), ballistic, or decelerating due to atmospheric drag.

Last, during interceptor support, the radar must provide the interceptor missile with data for commit or launch and guidance data.

12.2 MISSILE DEFENSE MISSION PARAMETERS AND REQUIREMENTS

BMD mission parameters vary with the type of threat, for example:

- Shorter-range TBMs (or submarine-launched ballistic missiles or SLBMs)

- Medium-range IRBMs (or submarine-launched ballistic missiles or SLBMs)

- Long-range ICBMs.

Key attributes for these three variations are the target dynamics and range of flight, the desired radar detection range and search coverage required to maximize battlespace or defended areas of the BMD system. The ranges of parameters are approximately:

- TBM defense (tactical BMD):
  - Range: 50 to 1,000 km
  - Target type: unitary and multistage
  - Threat range and speed: 200 to 1,000 km; 1,500 to 2,500 m/s.
- IRBM defense:
  - Range: 800 to 3,000 km
  - Target type: multistageThreat range and speed: 1,500 to 4,000 km; 2,000 to 6,000 m/s.
- ICBM defense (strategic BMD):
  - Range: 1,500 to 5,000 km
  - Target type: multistage
  - Threat range and speed: 4,000 to 12,000 km; 6,000 to 9,000 m/s.

The BMD radars must operate in the above environments, at various slant ranges, and against missile objects that vary in mean radar cross section (RCS). These widely varying conditions strongly influence the radar requirements.

12.3 INTERCEPTOR CAPABILITIES AND SUPPORT REQUIREMENTS

There are basically three categories of BMD interceptors:

- Short-range, endo-atmospheric
- Medium-range, endo-atmospheric or exo-atmospheric
- Long-range, exo-atmospheric.

Examples of these three categories are:

• Short-range: Patriot PAC-2 & PAC-3, Standard Missile-6 (SM-6)

• Medium-range: THAAD

• Long-range: EKV.

12.4   DEFENDED AREA

In general, the longer range BMD systems typically provide the largest bat-tlespace or defended areas (or keep-out zones) due to their longer ranges and faster interceptors. By negating the threats earlier in the timeline, a larger de-fended "footprint" is afforded by enhancing the keep-out or threat negation re-gion. The defended area size is driven by the radar, interceptor, or both, depending on the design balance achieved. Optimally, the radar should be sized to complement the interceptor.

12.5   BMD RADAR REQUIREMENTS

BMD radars requirements can be summarized in the following functional areas:

• Operating frequency:
  – C-band to X-band
• Antenna types:
  – Electronically steered arrays
  – Wideband time-delay steered phased-arrays
  – Possible mechanical and electronic steering
• Search types:
  – Autonomous horizon fence
  – Cued search (handovers from other surface, air, and space-based sensors)
• Tracking capability:
  – Track boosting, ballistic, and decelerating objects

- Tracking rates: variable
- Accuracy: Sufficient to support target classification algorithms
- Maneuver-tracking capability

• Target features:

- Signature
- Kinematics

• Classifier types:

- Bayes'
- Dempster-Shafer
- Decision trees

• Waveform characteristics:

- Single- and multiple-pulse (coherent integration)
- Search: Narrow bandwidth (500 kHz to 2 MHz)
- Track: Medium bandwidth
- Target classification: Narrow and wide bandwidths.

An example of tactical and strategic BMD parameters is provided in Table 12.1.

Table 12.1   BMD Radar Characteristics

Radar Parameter	Tactical Radar	Strategic Radar
Operating frequency	X-band	X-band
Antenna type	Electronically steered array (ESA)	Electronically steered array (ESA)
Search type	Autonomous horizon fence	Cued search
Tracking	Variable track rate	Variable track rate
Target features	Kinematics, signatures	Kinematics, signatures
Classifier type	Bayes'	Bayes'
Waveform characteristics	Narrow & wideband linear frequency modulation (LFM)	Narrow & wideband linear frequency modulation (LFM)

12.6   PERFORMANCE EVALUATION AND DESIGN VALIDATION

Evaluation of BMD radar performance can be performed at many levels of fidelity, including:

- Back-of-the-envelope analysis:
  - Radar range equation, K-factor analysis
- Desktop computer-aided analysis:
  - MATLAB or similar
  - Static and dynamic scenarios
  - Closed-form target fluctuation models (e.g., Swerling, log-normal)
  - Interceptor fly-out curves or equivalent
  - Single-run and Monte Carlo analyses
- High-fidelity simulation:
  - Detailed dynamic scenarios
  - Accurate target scattering models and dynamics
  - Waveforms and signal processing
  - Tracking filter and data association algorithms
  - Detailed feature modeling and classifier algorithms
  - Interceptor dynamics and fly-out models:
      Accelerations and burn-out velocities
      Divert capability
      Seeker models
- Real-time simulation:
  - Digital and/or hardware-in-the-loop
  - Actual real-time mission software
  - Monte Carlo trials
  - High-fidelity target simulation
      Radio frequency scattering models
      Data recording and reduction tools.

These performance assessments support radar systems analyses and design efforts during several phases of development, integration, and test, including:

- Radar architecture trade studies
- Radar system design trade-offs
- Subsystem requirements allocations
  - Hardware
  - Software
  - Interfaces
- Test vector generation for hardware and software design support
- Test plan and procedure development.

12.7 REFERENCES

[1] J. V. Candy, *Signal Processing—The Modern Approach*, McGraw-Hill
[2] S. Haykin & A. Steinhardt, *Adaptive Radar Detection and Estimation*, Wiley, 1992
[3] S. Haykin, *Adaptive Radar Signal Processing*, Wiley-Interscience, 2006
[4] S. Kay, *Modern Spectral Estimation: Theory and Application*, Prentice-Hall, 1999
[5] D. Manolakis, *Statistical and Adaptive Signal Processing*, Artech House, 2005
[6] S. L. Marple, *Digital Spectral Analysis with Applications*, Prentice-Hall, 1987
[7] R. A. Monzingo & T. M. Miller, *Introduction to Adaptive Arrays*, SciTech, 2003
[8] R. Nitzberg, *Radar Signal Processing and Adaptive Systems*, 2nd Edition, Artech House, 1999
[9] A. Oppenheim & R. Shafer, *Digital Signal Processing*, Prentice-Hall, 1975
[10] A. Papoulis, *Probability, Random Variables, and Stochastic Processes*, McGraw-Hill, 1965
[11] A. Papoulis, *Signal Analysis*, McGraw-Hill, 1977
[12] H. Van Trees, *Detection, Estimation and Modulation Theory, Part 1*, Wiley-Interscience, 2001
[13] Y. Bar-Shalom, *Multitarget-Multisensor Tracking: Principles and Techniques*, YBS, 1995
[14] Y. Bar-Shalom, *Multitarget/Multisensor Tracking: Applications and Advances*, Artech House, 2000
[15] S. Blackman & R. Popoli, *Design and Analysis of Modern Tracking Systems*, Artech House, 1999
[16] R. Duda, et al, *Pattern Classification*, 2nd Edition, Wiley-Interscience, 2000
[17] K. Fukunaga, *Introduction to Statistical Pattern Recognition*, 2nd Edition, Academic Press, 1990
[18] S. Theodoridis & K. Koutroumbas, *Pattern Recognition*, 2nd Edition, Academic Press, 2003

13

Early Warning Radar Design Considerations

13.1 INTRODUCTION

This chapter deals with some aspects of early warning radar design. Topics covered include:

- Early warning mission parameters and requirements
- Target/threat types:
  - Missile systems
  - Air targets
- Desired surveillance and associated functions:
  - Search requirements:
    - Missile target coverage
    - Air target coverage
  - Tracking requirements
  - Target classification and identification
  - Waveform requirements
- Performance evaluation and design validation.

Missile early warning radars (EWRs) are surveillance radars where the targets to be detected, tracked, and evaluated are ballistic missiles, including tactical ballistic missiles (TBMs), intermediate range ballistic missiles (IRBMs), or inter-continental ballistic missiles (ICBMs). Figure 13.1 illustrates the basic mission of an EWR. Steps A through E are typical functions for a missile EWR.

An EWR typically performs the search, acquisition, tracking, threat warning, and attack assessment functions. As such and due to the specific characteristics of ballistic missile targets, which are the primary threat type, these functions are specialized for this type of threat.

For example, many EWRs perform an autonomous horizon search fence to initially detect and acquire missiles. As described in Chapter 12, the horizon search fence is a resource-effective form of search that can be used to acquire ballistic missiles. As it avoids performing a large brute-force volume search, the EWR can be sized more modestly than, say, an anti-air warfare (AAW) or air defense (AD) radar, which must consider threats at varying ranges and altitudes over a large volume of space. Assuming that the radar is appropriately sized for the threats it must operate against, no missile can over-fly a horizon fence. Again, as shown in Chapter 1, there is a special form of the radar range equation for a horizon fence search that weakly favors lower operating frequencies.

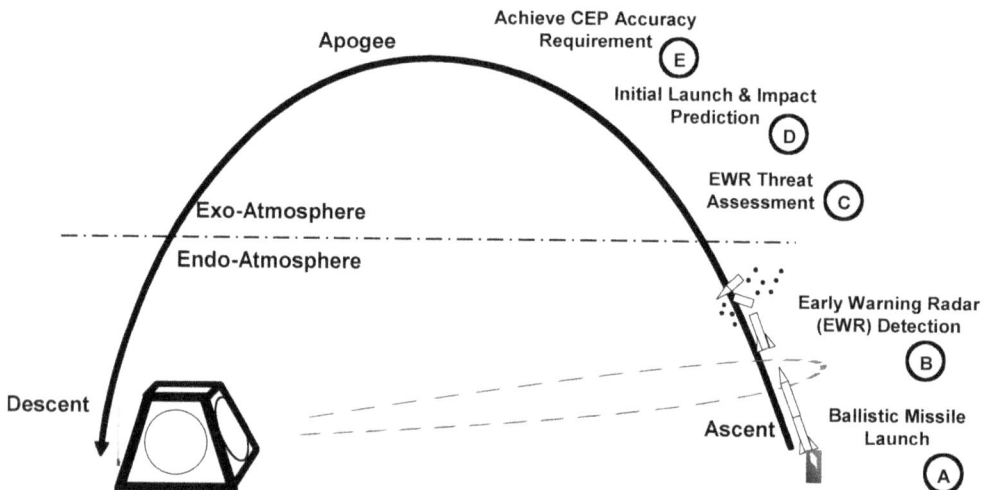

Figure 13.1 Mission of an Early Warning Radar

Tracking of ballistic missiles differs from other types of targets in that long-range missiles have three distinct phases of flight:

- Ascent or boost phase (rocket booster is accelerating while within the earth's atmosphere)

- Midcourse or ballistic phase

- Descent or reentry (missile objects reenter Earth's atmosphere and are subject to drag deceleration).

For most longer-range EWR applications, horizon fence detection will occur after the missiles are ballistic (due to the radar horizon). Some shorter-range applications can observe missiles during the boost phase. Tracking filters, therefore, must operate over one or more of these flight regimes, depending on the role of the particular EWR system.

Target classification and discrimination of ballistic missiles during the above phases of flight can use different approaches. Single and multiple objects need to be classified depending on the phase of flight.

The threat assessment and attack warning functions use tracking data and simple kinematics and signature features to distinguish missile targets from air targets or satellites. Bayes' techniques can be used for this purpose.

13.2 EARLY WARNING MISSION PARAMETERS AND REQUIREMENTS

Early warning (EW) mission parameters vary with the type of threat, for example:

- Shorter-range TBMs (or submarine-launched ballistic missiles or SLBMs)

- Medium-range IRBMs (or submarine-launched ballistic missiles or SLBMs)

- Long-range ICBMs.

Key attributes for these three variations are the target dynamics and range of flight, the desired radar detection range and search coverage required to maximize the threat assessment and attack warning regions of the EW system. The ranges of parameters are approximately:

- TBM EW (tactical EW):

  - Range: 50 to 1,000 km
  - Target type: unitary and multistage
  - Threat range and speed: 200 to 1,000 km; 1,500 to 2,500 m/s

- IRBM EW:

  - Range: 800 to 3,000 km
  - Target type: multistage
  - Threat range and speed: 1,500 to 4,000 km; 2,000 to 6,000 m/s

- ICBM defense (strategic EW):

  - Range: 1,500 to 5,000 km
  - Target type: multistage
  - Threat range and speed: 4,000 to 12,000 km; 6,000 to 9,000 m/s.

The EWRs must operate in the above environments, at various slant ranges, and against missile objects that vary in mean radar cross section (RCS). As for BMD radars, these widely varying conditions strongly influence EWR requirements.

13.3 THREAT WARNING AND ATTACK ASSESSMENT

In general, the longer-range EW systems typically provide the largest threat warning and attack assessment regions. In terms of the types of "assets" warned by the different EW systems, short-range systems provide relatively small coverage, such as for localized personnel and equipment, whereas a tactical EW can warn entire cities or larger regions, and strategic EW systems can provide warning for entire countries.

13.4 EWR REQUIREMENTS

Typical EWR radar requirements can be summarized in the following functional areas:

- Typical operating frequency:
  - UHF to L-band
- Antenna types:
  - Full-FOV (FFOV)
  - Narrow bandwidth phased arrays
- Search types:
  - Autonomous horizon fence
  - Cued search for satellite acquisition
- Tracking capability:
  - Track boosting, ballistic, and decelerating objects
  - Tracking rates: variable
  - Accuracy: Sufficient to satisfy warning requirements
  - May track air and surface targets
- Target features for missile versus aircraft and satellites:
  - Signature
  - Kinematics
- Classifier types:
  - Bayes' or decision trees
- Waveform characteristics:
  - Single and multiple-pulse (coherent integration)
  - Search: Narrow bandwidth (500 kHz to 2 MHz)
  - Track and classification: Medium bandwidth.

An example of tactical and strategic EWR parameters is provided in Table 13.1.

13.5 PERFORMANCE EVALUATION AND DESIGN VALIDATION

Evaluation of EWR performance can be performed at many levels of fidelity, including:

Table 13.1 EWR Characteristics

Radar Parameter	Tactical EW	Strategic EW
Operating frequency	UHF, L-band	UHF, L-band
Antenna type	FFOV	FFOV
Search type	Autonomous horizon fence	Autonomous horizon fence
Tracking	Track rate: variable	Track rate: variable
Target features	Kinematics, signatures	Kinematics, signatures
Classifier type	Bayes' or decision tree	Bayes' or decision tree
Waveform characteristics	Narrow bandwidth LFM	Narrow bandwidth LFM

- Back-of-the-envelope analysis:

  - Radar range equation, Sorensen-type tracking analysis

- Desktop computer-aided analysis:

  - MATLAB or similar
  - Static and dynamic scenarios
  - Closed-form target fluctuation models (e.g., Swerling, log-normal)
  - Single-run and Monte Carlo analyses

- High-fidelity simulation:

  - Detailed dynamic scenarios
  - Waveforms and signal processing
  - Tracking filter, threat assessment, and attack warning algorithms

- Real-time simulation:

  - Digital and/or hardware-in-the-loop
  - Actual real-time mission software
  - Monte Carlo trials
  - High-fidelity target simulation
        Radio frequency scattering models
        Data recording and reduction tools.

These performance assessments support radar systems analyses and design efforts during several phases of development, integration, and test, including:

- Radar architecture trade studies

- Radar system design trade-offs

- Subsystem requirements allocations

    – Hardware
    – Software
    – Interfaces

- Test vector generation for hardware and software design support

- Test plan and procedure development.

13.6 REFERENCES

[1] J. V. Candy, *Signal Processing—The Modern Approach*, McGraw-Hill,

[2] S. Haykin & A. Steinhardt, *Adaptive Radar Detection and Estimation*, Wiley, 1992

[3] S. Haykin, *Adaptive Radar Signal Processing*, Wiley-Interscience, 2006

[4] S. Kay, *Modern Spectral Estimation: Theory and Application*, Prentice-Hall, 1999

[5] D. Manolakis, *Statistical and Adaptive Signal Processing*, Artech House, 2005

[6] S. L. Marple, *Digital Spectral Analysis with Applications*, Prentice-Hall, 1987

[7] R. A. Monzingo & T. M. Miller, *Introduction to Adaptive Arrays*, SciTech, 2003

[8] R. Nitzberg, *Radar Signal Processing and Adaptive Systems*, 2nd Edition, Artech House, 1999

[9] A. Oppenheim & R. Shafer, *Digital Signal Processing*, Prentice-Hall, 1975

[10] A. Papoulis, *Probability, Random Variables, and Stochastic Processes*, McGraw-Hill, 1965

[11] A. Papoulis, *Signal Analysis*, McGraw-Hill, 1977

[12] H. Van Trees, *Detection, Estimation and Modulation Theory, Part 1*, Wiley-Interscience, 2001

[13] Y. Bar-Shalom, *Multitarget-Multisensor Tracking: Principles and Techniques*, YBS, 1995

[14] Y. Bar-Shalom, *Multitarget/Multisensor Tracking: Applications and Advances*, Artech House, 2000

[15] S. Blackman & R. Popoli, *Design and Analysis of Modern Tracking Systems*, Artech House, 1999

[16] R. Duda, et al., *Pattern Classification*, 2nd Edition, Wiley-Interscience, 2000

[17] K. Fukunaga, *Introduction to Statistical Pattern Recognition*, 2nd Edition, Academic Press, 1990

[18] S. Theodoridis & K. Koutroumbas, *Pattern Recognition*, 2nd Edition, Academic Press, 2003

14

Air Defense Radar Design Considerations

14.1 INTRODUCTION

This chapter discusses some of the key aspects of air defense radar design. Topics covered include:

• Air defense mission parameters and requirements

• Air target threat types:

 – Aircraft, unmanned aerial vehicles (UAVs), cruise missiles
 – Chaff and jamming

• Interceptor capabilities:

 – Maximum velocity
 – Fly-out range
 – Interceptor support requirements

• Desired defended area:

 – Search requirements

- – Tracking requirements
- – Target features
- – Classifier requirements
- – Waveform requirements

• Performance evaluation and design validation.

Air defense (AD) radars are fire control radars where the intended target is an air target, including aircraft, helicopters, UAVs, and cruise missiles. Figure 14.1 illustrates the operation of an AD radar.

An AD radar typically performs the search, acquisition, tracking, target classification and discrimination, and interceptor support functions. As such and due to the widely varying characteristics of air targets, each of these functions must accommodate a wide range of threats. Shipboard AD systems are often referred to as anti-air warfare (AAW) systems.

For example, most AD radars perform some form of a volume search to initially detect and acquire targets. In addition, they can be cued by other sensors (e.g., a handover from another radar). Assuming that the AD radar is appropriately sized for the threats it must operate against, all targets will be detected in the volume search. As shown in Chapter 1, there is a special form of the radar range equation for a volume search that is independent of radar operating frequency.

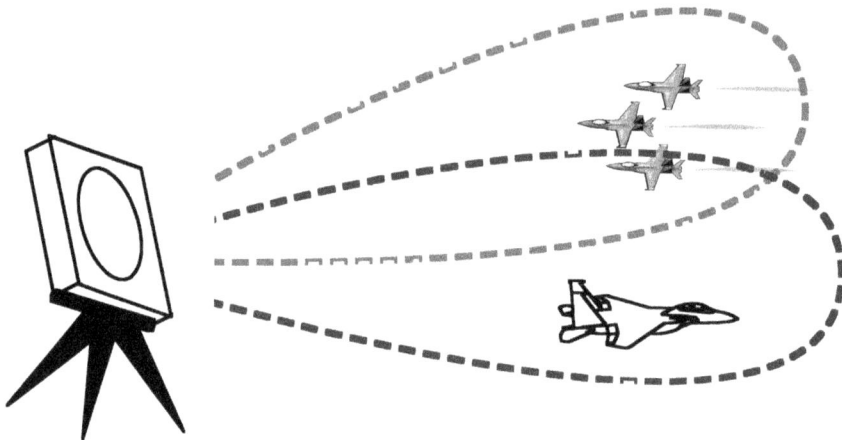

Figure 14.1 Mission of Air Defense Radar

Tracking of air targets differs from other types of targets in that these targets can exhibit many distinct types of flight:

- Altitudes from "on the deck" or about 15 meters to 80,000 feet

- May be manned or unmanned (i.e., pull low and very high-g maneuvers, and be unpredictable in doing so)

- Straight and level, constant velocity

- Straight and level, linear acceleration

- High-g maneuvers (turns, climbs, dives)

- Near-stationary or hovering

- Very-low altitude, terrain-following

- May employ propellers, jets, and rockets.

Tracking filters, therefore, must operate over many or all of these flight regimes and target dynamics, depending on the role of the particular AD system. In many cases where many types of targets and associated dynamics must be accommodated, multiple-model tracking filters, such as the interacting multiple-model (IMM) filter may be appropriate. For slower air targets, or surface targets (included here for completeness), track-while-scan (TWS) methods can often be employed.

Target classification and discrimination of air threats during the above phases of flight is referred to by the term non-cooperative target recognition (NCTR). In addition to NCTR, some systems attempt to perform identification (ID).

Last, during interceptor support, the radar must provide the interceptor missile with data for commit or launch, guidance data or target illumination, depending on the type of interceptor being employed by the AD system.

14.2   AIR DEFENSE MISSION PARAMETERS AND REQUIREMENTS

AD mission parameters vary with the type of threat, for example:

- Threat aircraft in presence of possible friendly aircraft

- Cruise missile (CM) or anti-radiation missile (ARM) attacks
- Slowly-moving threats, e.g., helicopters or UAVs.

Key attributes for these variations are the target dynamics and range of flight, the desired radar detection range and search coverage required to maximize defended areas of the AD system. The ranges of parameters are approximately:

- Tactical air defense:
  - Range: 10 to 200 km
  - Target type: aircraft, CMs, helicopters, UAVs
  - Threat range and speed: 10 to 200 km; 50 m/s to 2,000 m/s.
- Strategic air defense:
  - Range: 50 to 500 km
  - Target type: aircraft, CMs, helicopters, UAVs
  - Threat range and speed: 50 to 500 km; 50 m/s to 2,000 m/s.

The AD radars must operate in the above environments, at various slant ranges, and against air threats that vary in mean radar cross section (RCS) . These widely varying conditions strongly influence the radar requirements.

14.3 INTERCEPTOR CAPABILITIES AND SUPPORT REQUIREMENTS

There are basically three categories of interceptors:

- Short-range guns and missiles
- Medium-range missiles
- Long-range missiles.

Some approximate capabilities are:

- Short-range gun
  - Speed: mach 3
  - Range: 5 to 10 km
- Medium-range missile:
  - Speed: mach 3.5
  - Range: 50+ km

- Long-range missile:
  - Speed: mach 4+
  - Range: 75+ km.

14.4 DEFENDED AREA

In general, the longer-range AD systems typically provide the largest battlespace or defended areas (or keep-out zones) since due to their longer ranges and faster interceptors, by negating the threats earlier in the timeline, a larger defended "footprint" is afforded by enhancing the keep-out or threat negation region. In terms of the types of "assets" defended by the different AD systems, short-range systems like guns provide relatively small coverage, such as for localized personnel and equipment or ships, whereas a missile can defend against medium-range aircraft and CMs, and longer-range systems can provide defense of larger areas, such as a convoy of ships or small cities, against both air and even some missile threats. The defended area size is driven by the radar, interceptor, or both, depending on the design balance achieved.

14.5 AIR DEFENSE RADAR REQUIREMENTS

AD radars requirements can be summarized in the following functional areas:

- Typical operating frequency:
  - S-band to C-band
- Antenna types:
  - Full FOV (FFOV)
  - Narrow bandwidth phased arrays
- Search types:
  - Volume search
  - Autonomous horizon fence
  - Cued search (handovers from other surface or airborne sensors)
- Tracking capability:
  - Track manned and unmanned
  - Tracking rates: medium, low (straight and level)

- Accuracy: Sufficient to support weapon requirements
- Maneuver-tracking capability

• Waveform characteristics:

- Single and multiple-pulse (coherent integration)
- Pulse-Doppler
- Search: Narrow bandwidth
- Track: Medium bandwidth.

Typical AD parameters are provided in Table 14.1.

Table 14.1   AD Radar Characteristics

Radar Parameter	Radar Attribute
Operating frequency	S-band, C-band
Antenna type	FFOV
Search type	Volume, autonomous horizon fence
Tracking	Track rate: variable
Waveform characteristics	LFM: various bandwidths

14.6   PERFORMANCE EVALUATION AND DESIGN VALIDATION

Evaluation of AD radar performance can be performed at many levels of fidelity, including:

• Back-of-the-envelope analysis:

- Radar range equation, Sorensen-type tracking analysis

• Desktop computer-aided analysis:

- MATLAB or similar
- Static and dynamic scenarios
- Closed-form target fluctuation models (e.g., Swerling, log-normal)
- Interceptor fly-out curves or equivalent
- Single-run and Monte Carlo analyses

- High-fidelity simulation:
  - Detailed dynamic scenarios
  - Accurate target scattering models and dynamics
  - Waveforms and signal processing
  - Tracking filter and data association algorithms
  - Detailed feature modeling and classifier algorithms
  - Interceptor dynamics and fly-out models:
      Accelerations and burn-out velocities
      Seeker models

- Real-time simulation:
  - Digital and/or hardware-in-the-loop
  - Actual real-time mission software
  - Monte Carlo trials
  - High-fidelity target simulation
      Radio frequency scattering models
      Data recording and reduction tools.

These performance assessments support radar systems analyses and design efforts during several phases of development, integration, and test, including:

- Radar architecture trade studies

- Radar system design trade-offs

- Subsystem requirements allocations:
  - Hardware
  - Software
  - Interfaces

- Test vector generation for hardware and software design support

- Test plan and procedure development.

14.7 REFERENCES

[1] J. V. Candy, *Signal Processing—The Modern Approach,* McGraw-Hill, 1989
[2] S. Haykin & A. Steinhardt, *Adaptive Radar Detection and Estimation,* Wiley, 1992
[3] S. Haykin, *Adaptive Radar Signal Processing,* Wiley-Interscience, 2006

[4] S. Kay, *Modern Spectral Estimation: Theory and Application,* Prentice-Hall, 1999

[5] D. Manolakis, *Statistical and Adaptive Signal Processing,* Artech House, 2005

[6] S. L. Marple, *Digital Spectral Analysis with Applications,* Prentice-Hall, 1987

[7] R. A. Monzingo & T. M. Miller, *Introduction to Adaptive Arrays,* SciTech, 2003

[8] R. Nitzberg, *Radar Signal Processing and Adaptive Systems,* 2nd Edition, Artech House, 1999

[9] A. Oppenheim & R. Shafer, *Digital Signal Processing,* Prentice-Hall, 1975

[10] A. Papoulis, *Probability, Random Variables, and Stochastic Processes,* McGraw-Hill, 1965

[11] A. Papoulis, *Signal Analysis,* McGraw-Hill, 1977

[12] H. Van Trees, *Detection, Estimation and Modulation Theory, Part 1,* Wiley-Interscience, 2001

[13] Y. Bar-Shalom, *Multitarget-Multisensor Tracking: Principles and Techniques,* YBS, 1995

[14] Y. Bar-Shalom, *Multitarget/Multisensor Tracking: Applications and Advances,* Artech House, 2000

[15] S. Blackman & R. Popoli, *Design and Analysis of Modern Tracking Systems,* Artech House, 1999

[16] R. Duda, et al., *Pattern Classification,* 2nd Edition, Wiley-Interscience, 2000

[17] K. Fukunaga, *Introduction to Statistical Pattern Recognition,* 2nd Edition, Academic Press, 1990

[18] S. Theodoridis & K. Koutroumbas, *Pattern Recognition,* 2nd Edition, Academic Press, 2003

15

Predicted Performance of Phased-Array Radars

15.1 INTRODUCTION

The performance evaluation of phased-array radars is addressed in this chapter. Topics covered include:

- Functional performance:
  - Target detection:
    - Clear environments
    - Clutter environments
    - Jamming environments
  - Tracking:
    - Ballistic missile targets
    - Air targets
  - Interference suppression performance:
    - Sidelobe canceller performance
    - Open-loop nulling performance
    - Adaptive array performance
    - Doppler processing performance
    - Time and frequency domain excision performance

- Clutter cancellation performance:
    Land clutter
    Sea clutter
    Rain and other weather clutter
- Hardware subsystems:
    Range sidelobes
    Channel-to-channel alignment
    Amplitude and phase errors
    Phase noise evaluation
    Wideband operation versus subarray size.

Two basic types of performance prediction are discussed: (1) functional performance, that is, how well a specified radar performs its functions (e.g., search, track) and (2) mission-level performance, that is, how well the radar performs its mission (e.g., air defense, missile defense, early warning).

The following sections address the specific aspects of functional and mission-level performance prediction.

15.2 FUNCTIONAL PERFORMANCE

15.2.1 Target Detection

Ultimately, predicting target detection performance entails calculating the probability of detection (P_D) for a specified target, range, radar cross section (RCS), and waveform for a defined probability of false alarm (P_{FA}). The detailed approach to accomplishing this objective will vary with the particular type of target model, detection rule, waveform parameters, and signal-to-noise ratio (SNR). The theoretical bases for this performance prediction are provided in Chapter 2 and, for example, [12].

Consider a ± 45-degree azimuth horizon search fence for ballistic missile acquisition that covers from 500 km out to 1,500 km. The missile RCS follows a Swerling I fluctuation model, exhibits a mean RCS of 0 dBsm, and develops signal-to-noise ratios of 10, 15, and 20 dB depending on radar sizing. If the search is designed to provide four statistically independent looks at a missile while in the fence, then the cumulative probability of detection can be calculated. An alternative design that must operate in narrowband sidelobe noise jamming uses frequency hopping to avoid the jammer. Assume the frequency hops are sufficiently diverse in frequency to decorrelate the target from look to look. This has the result of transforming the Swerling I fluctuation model into a Swerling II model. The predicted performance of the two alternative designs is to be calculated. Table 15.1 displays the results.

Table 15.1   Horizon Search Fence with Swerling I and II Targets

SNR (dB)	Swerling I		Swerling II	
	Single-Pulse Probability of Detection	Cumulative Probability of Detection for 4 Independent Looks	Single-Pulse Probability of Detection	Cumulative Probability of Detection for 4 Independent Looks
10	0.285	0.738	0.091	0.318
15	0.655	0.986	0.401	0.871
20	0.872	1.000	0.735	0.995

Comparing the cumulative probabilities of detection for the two Swerling models indicates that if a cumulative probability of at least 0.96 is required, then with no jamming (i.e., Swerling I) an SNR of 15 dB provides margin, whereas for the jamming case with hopping (i.e., Swerling II), closer to 20 dB SNR is required. One interpretation of these results is that the jammer degrades search performance by nearly 5 dB (or alternatively, an approximate 5 dB larger radar is required for equivalent performance). A similar analysis could be performed for non-coherent integration, with the results compared to those for binary integration case above.

Table 15.2 considers an alternative design with sidelobe cancellers that provide an average of 15 dB of jammer cancellation that achieves signal-to-interference ratios (SIRs) of 9, 13.5, and 18 dB. As can be seen, by incorporating sidelobe cancellers into the radar system design, a radar that in the clear produces 15 dB SNR, will achieve 13.5 dB SIR and meets the cumulative probability of detection requirement of 0.96 (with no margin).

Now consider the same radar operating in a severe clutter environment such that the signal-to-clutter ratios (SCRs) are –23 dB, –18 dB, and –13 dB, respectively. If a three-pulse moving target indicator (MTI) canceller provides clutter mitigation adequate to produce 8.5 dB, 13 dB, and 17.5 dB, respectively, then Table 15.3 contains the corresponding performance predictions.

Table 15.2   Horizon Search Fence with Swerling I Targets and Sidelobe Cancellers

SIR (dB)	Single-Pulse Probability of Detection	Cumulative Probability of Detection for 4 Independent Looks
9	0.213	0.617
13.5	0.554	0.960
18	0.806	0.999

Table 15.3   Horizon Search Fence with Swerling I Targets in a Clutter Environment with MTI Cancellation

Input SCR (dB)	Cumulative Probability of Detection for 4 Independent Looks	SCR after MTI Cancellation (dB)	Cumulative Probability of Detection for 4 Independent Looks
−23	0.181	8.5	0.550
−18	0.517	13.0	0.946
−13	0.786	17.5	0.998

As can be seen from the table, the radar that in the clear produces 15 dB SNR, produces 13 dB SCR after MTI cancellation and achieves a cumulative probability of detection of 0.946. Although not quite equal to the performance in the clear, this predicted performance indicates very little degradation and would probably be considered adequate in most situations.

15.2.2   Tracking

Tracking performance predictions usually focus on track smoothing and prediction accuracies. Two target types will be considered: (i) aircraft with constant velocity, and (ii) intact tactical ballistic missile (TBM) after rocket burnout. Assume the radar achieves 15 dB SNR on both targets and has a 3 dB beamwidth of 2 degrees. The aircraft is tracked at 5 Hz and the missile is tracked at 2 Hz. A time-in-track of 10 seconds is achieved on the aircraft prior to a maneuver, and a time-in-track of 30 seconds is achieved for the missile target. Both targets are 250 km from the radar. The track smoothing estimation error is approximately given by:

$$\sigma_P = \frac{R\,\theta_3}{\sqrt{2\,SNR\,f_r\,T_t}}\,, \tag{15.1}$$

where R, θ_3, SNR, f_r, and T_t are the target average slant range, antenna 3 dB beamwidth, signal-to-noise ratio, tracking rate, and time-in-track, respectively. In addition, the predicted position error is approximately:

$$\sigma_{P-Predicted} = \frac{R\theta_3}{\sqrt{2\,SNR\,f_r\,T_t}} + \frac{T_P\,\sqrt{12}\,R\theta_3}{T_t\,\sqrt{2\,SNR\,f_r\,T_t}}\,. \tag{15.2}$$

The position is to be predicted 10 seconds ahead (T_P). Table 15.4 illustrates the predicted performance for this tracking problem.

Table 15.4 Smoothed and Predicted Position Errors for Aircraft and Missile

Target types	One-Sigma Smoothed Position Error (m)	One-Sigma Predicted Position Error (m)
Aircraft	157	693
TBM	143	305

It should be noted that the longer time-in-track for the TBM (i.e., 30 seconds versus 10 seconds) outweighs the higher track rate for the aircraft (5 Hz for the aircraft, while the TBM's trajectory, if outside the atmosphere, is effectively ballistic acting only under gravity, versus 2 Hz for the TBM) for equal SNR. This is to be expected. In reality performance could favor TBM tracking even more since aircraft inevitably maneuver, which would further limit the smoothing time-in-track for the aircraft. However, a larger SNR would probably result from the aircraft as compared with the TBM.

15.2.3 Interference Suppression

Prediction of the interference mitigation performance of phased-arrays radar can be performed at a first-pass, or "back-of-the-envelope," by using approximate formulas, as well as by using high-fidelity simulations.

Sidelobe canceller performance is addressed in Chapter 8. The residual jammer power after an ideal sidelobe canceller is, from Chapter 8 and [8], given by:

$$for \ P_J >> P_n \ and \ \rho = 1 \qquad P_{min} \approx P_n \left[1 + \frac{G_m}{G_a} \right] \qquad (15.3)$$

where P_J, P_n, ρ, G_m, and G_a are the jammer and thermal noise power, correlation of the jammer in the main and auxiliary channels, and the main antenna sidelobe and auxiliary antenna gains, respectively. Table 15.5 evaluates equation (15.9) for several antenna gain values when the SNR in the clear is 20 dB to calculate output signal-to-interference ratio.

Table 15.5   Output SIR (dB) Using an Ideal Sidelobe Canceller

G_m/G_a (dB)	Output SIR (dB)
3	15.22
0	16.99
−10	19.59
−20	19.96

As can be seen, good performance is achieved when the auxiliary antenna gain is at least 10 dB higher than the main antenna sidelobes. However, even with the sidelobe canceller, the radar's performance is still degraded by 3 dB for auxiliary to sidelobe gain of unity. For a Swerling I target and a probability of false alarm of 10^{-6}, this results in a significant loss in target detectability: probability of detection of 0.87 in the clear versus 0.76 for operating in jamming with a low-gain auxiliary antenna when using a sidelobe canceller.

Open-loop nulling can be employed to cancel interference by estimating the optimal weight vector using a number of interference samples. Consider an example where the ideal mitigation using the optimal weight is −35 dB cancellation of the interference. The open-loop nulling performance for the case of a 32-element array is estimated versus number of samples used to compute the weights with the approximate results appear in Table 15.6. The calculation uses

Table 15.6   Open-Loop Nulling Performance vs. Number of Samples used to Compute the Weights for an Ideal –35 dB Cancellation

Number of Samples	Open-Loop Cancellation (dB)
32	–22.82
48	–30.65
64	–32.18
80	–32.90
96	–33.32
128	–33.81
160	–34.07

equation (15.4) for the approximate loss relative to the ideal cancellation, where N and M are the number of samples used and number of antenna elements, respectively, from Chapter 8 and [8] given by:

$$Loss\ (dB)\ =\ 10\ log_{10}\left[\frac{N+2-M}{N+1}\right].\tag{15.4}$$

As can be seen in the table, open-loop cancellation is significantly degraded when the number of samples equals the number of elements (i.e., 12 dB), versus 80 samples (i.e., 2 dB) or 160 samples (i.e., 1 dB). The trade-off is the time required to collect the interference samples, during which the interference parameters could change resulting in a loss in cancellation performance.

Similar performance can be achieved for open-loop Doppler nulling of clutter backscatter using a limited number of clutter samples. Closed-loop interference or clutter mitigation can often achieve superior cancellation performance compared to open-loop nulling, at the cost of higher computational requirements. Again, estimating the adaptive weights using a limited number of interference or clutter samples will suffer a similar loss in performance as predicted by equation (15.4).

Another type of interference suppression is the use of time or frequency-domain censoring in the signal processor. This can be extremely effective when interference (or clutter backscatter) is of short duration or is narrowband in nature. Equivalently, for clutter suppression, this corresponds to discrete clutter returns like those arising from man-made structures. The performance is limited by the number of interference-dominated samples versus the total number of signal-plus-interference samples. Table 15.7 illustrates the predicted loss in performance versus the ratio of censored samples to total samples (e.g., range cells, frequency bins, or Doppler filters). As seen in the table, up to approximately 10% of samples can be censored with an acceptable loss when compared with the degradation of uncensored interference sources. In some cases, accepting a 1 or 2 dB loss would also be acceptable.

Table 15.7   Loss in SNR due to Time or Frequency Censoring as a Function of the Ratio of Censored Samples to Total Samples

Ratio of Censored Number of Samples to Total Number	Loss in SNR Due to Censoring (dB)
0.05	−0.22
0.10	−0.46
0.20	−0.96
0.30	−1.55
0.40	−2.22
0.50	−3.01

It should be noted that the cancellation performance discussed in this section assumes no amplitude or phase errors, nor channel-to-channel mismatch errors, that exist due to hardware subsystem designs. These effects are addressed in Section 2.6.

15.2.4   Clutter Cancellation

Clutter mitigation is usually performed via the signal processing techniques described in Chapter 2. The two types of mitigation are MTI waveforms and processing and pulse-Doppler waveforms and processing. The performances of both types of clutter cancellation are addressed in this section.

MTI processing is effective for zero Doppler-shifted clutter (or at most moving very slowly relative to targets of interest). This technique is a digital high-pass filter that has zero gain at zero Doppler and all multiples of the pulse repetition frequency (PRF), as depicted in Figure 2.7 of Chapter 2. For these conditions, good performance can be attained as approximately given for two and three-pulse cancellers by the relationships:

$$I_2 = CA \approx \left[2 \left(\pi \sigma_f T \right)^2 \right]^{-1} = \left(PRF / \sigma_f \right)^2 / 19.75 \qquad (15.5)$$

and

$$I_3 \approx \left(\frac{PRF}{\sigma_f} \right)^4 / 780, \qquad (15.6)$$

where σ_f, T, and PRF are the one-sigma clutter spectral width, the pulse repetition interval (PRI), and pulse repetition frequency, respectively. Table 15.8 illustrates the typical theoretical performance of the 2 and 3-pulse MTI cancellers.

Table 15.8   Predicted MTI Clutter Cancellation Performance

Ratio of PRF to clutter spectral width	2-Pulse MTI cancellation (dB)	3-Pulse MTI cancellation (dB)
0.05	−38.97	−80.96
0.10	−32.96	−68.92
0.25	−25.00	−53.00
0.40	−20.91	−44.84

As seen in Table 15.8 canceller performance degrades with increasing clutter spectral width. However, for relatively narrow clutter spectra the two-pulse MTI canceller provides good performance. The three-pulse MTI can mitigate wider-spectrum clutter. Note that these cancellation values will be degraded for non-zero velocity clutter and other real-world effects such as timing jitter or non-stationary clutter statistics.

Pulse-Doppler waveforms and processing provide superior performance to MTI, especially for nonzero velocity clutter, at the cost of more radar resources (e.g., 16 or 32 coherent pulses) and signal processing (i.e., both pulse matched filtering and processing for N Doppler filters, where N is the number of coherent pulses in the train).

For clutter spectra that are narrower than the Doppler filter bandwidth, near-perfect clutter cancellation can result. However, for broad clutter spectra, such as for blowing rain, more than one filter might have significant clutter. Elimination of all Doppler filters with clutter will eventually reduce the target Doppler coverage and will require a larger number of PRFs (and coherent batches of pulses) to regain adequate target velocity coverage. Therefore to minimize this effect, filters with some clutter are typically used for detection purposes, and will result in less clutter cancellation.

15.2.5 Hardware Subsystems

This section addresses some common hardware-limited performance issues.

15.2.5.1 Range (Time) Sidelobes.
Hardware errors affect achievable range sidelobe levels. Key contributors to this phenomenon are the signal paths in the antenna and receiver analog hardware. Approximate range sidelobe level is limited to no better than the root-sum-squared errors from the antenna, receiver, and sidelobe taper via:

$$SLL\ (dB)\ =\ 10\log_{10}\sqrt{Error_{ant}^2 + Error_{rec}^2 + Taper^2}\ . \tag{15.7}$$

Table 15.9 illustrates the achievable range sidelobes for reasonable antenna and receiver error magnitudes for range sidelobes of −30 dB and −40 dB relative to the peak response at the matched filter output. As can be seen, about 5 dB margin exists for the desired −30 dB sidelobes, and about 3.4 dB margin is available for the −40 dB desired sidelobe case. If the receiver and antenna errors rose to −42 dB, and a −45 dB taper was again selected, then the −40 dB sidelobe level would be met with no margin.

Table 15.9   Predicted Range Sidelobe Performance

Error and Taper Levels (dB)	SLL for −30 dB	SLL for −40 dB
Receiver: −50		
Antenna: −45	−34.97	−43.39
Taper: $SSL_{desired}$ − 5 dB		

15.2.5.2 Channel-to-Channel Alignment. Predicted interference suppression performance will be degraded when channel-to-channel mismatch arises. Chapter 11 addresses this and provides a formula for the achievable interference cancellation as a function of channel-to-channel alignment errors:

$$CR = \left(\frac{4}{3}\right)\left(\varepsilon_p - 1\right)^2. \qquad (15.8)$$

Consider the adaptive array that provides −35 dB cancellation of a sidelobe noise jammer. Table 15.10 shows the effects of various degrees of channel mismatch.

Table 15.10   Degradation of Cancellation as a Function of Channel Mismatch

Channel Mismatch Error (dB)	Cancellation Performance (dB)
0.02	−35.0
0.05	−35.0
0.10	−35.0
0.20	−31.4
0.30	−27.8
0.40	−25.3

As seen from Table 15.10, when more than 0.1 dB of mismatch error exists for this example, cancellation performance is degraded.

15.2.5.3 Amplitude and Phase Errors. Similar to the effect of the channel matching evaluated errors in Section 2.6.2, amplitude and phase errors will also degrade interference mitigation performance. In Chapter 11, the maximum cancellation versus sinusoidal amplitude error in the passband is given by:

$$Amplitude\ Error\,(dB)\ =\ 20\ \log_{10}\left[\,(1+\delta/2)/(1-\delta/2)\,\right],\qquad(15.9)$$

where δ is the peak amplitude of the sinusoid. As can be seen from Table 15.11, as for channel mismatch errors, for peak sinusoidal errors of greater than 0.1 dB, cancellation performance is degraded.

Table 15.11   Degradation of Cancellation as a Function of Amplitude Errors

Peak Sinusoidal Mismatch Error (dB)	Achievable Cancellation Ratio (dB)
0.02	−35.0
0.05	−35.0
0.10	−35.0
0.20	−32.7
0.30	−29.1
0.40	−25.3

For phase errors, a similar relationship relating to cancellation performance exists. In order to achieve the −35 dB cancellation ratio requires that phase errors be on the order of 5 degrees rms.

15.2.5.4 Phase Noise. Phase noise is the primary limitation to achievable clutter cancellation as discussed in Chapter 11. The calculation method used here follows the procedure outlined in [19]. This method is graphical in nature and effectively adjusts the receiver phase noise frequency response by first multiplying it by the system frequency response function (defined like a Bode plot in control systems design) and then integrating over frequency to obtain the resulting phase noise. The system frequency response is defined by the range to clutter, the type of MTI canceller used, and the waveform's pulse bandwidth.

First, to account for clutter cancellation occurring due to correlation between the transmit and receive (assuming a common local oscillator source), the break point frequency is computed as:

$$f_c = \frac{c}{2\pi R_c},$$ (15.10)

where c and R_c are the speed of light and mean slant range to the clutter, respectively. Consider the following example. For a mean range of 100 km, the clutter break frequency is approximately 500 Hz. The frequency response slope for this "leg" of the plot is 20 dB/decade.

Next, for the three-pulse MTI canceller, the approximate break frequency is computed as $f_2 = 0.249$ (PRF_{ave}), or about 119 Hz. The frequency response slope for this portion of the plot is 40 dB/decade. Last, the intermediate frequency filter response is assumed to have a ± 1.25 MHz 3 dB bandwidth (i.e., corresponding to a 2.5 MHz pulse bandwidth) and is assumed to roll off at –40 dB/decade above 1.25 MHz.

A nominal receiver phase noise frequency response is assumed with its floor at a level of –140 dBc/Hz above the break frequency of 1.75 kHz. A roll-off of –30 dB/decade is assumed below the break frequency. To account for the phase noise spectrum being two-sided, and noise being present on the oscillator for both transmit and receive, the assumed response is increased by 6 dB. The calculated system frequency response and the adjusted phase noise frequency response are plotted in Figure 15.1.

As indicated on Figure 15.1, the integrated phase noise is –62.6 dBc (i.e., dB relative to the carrier). Therefore, –62.6 dB is the predicted maximum clutter cancellation ratio. In general, 5 to 10 dB of margin would be allocated in the design, which would result in a clutter cancellation capability of approximately –55 dB, quite adequate for many applications. A similar analysis approach can be used for pulse-Doppler waveforms and processing.

15.2.5.5 Wideband Operation versus Subarray Size. Due to the frequency-steering phenomenon that exists for phased-arrays operated over very wide bandwidths, a loss referred to as dispersion loss results from the antenna main lobe scanning across a target versus instantaneous waveform frequency. For linear frequency

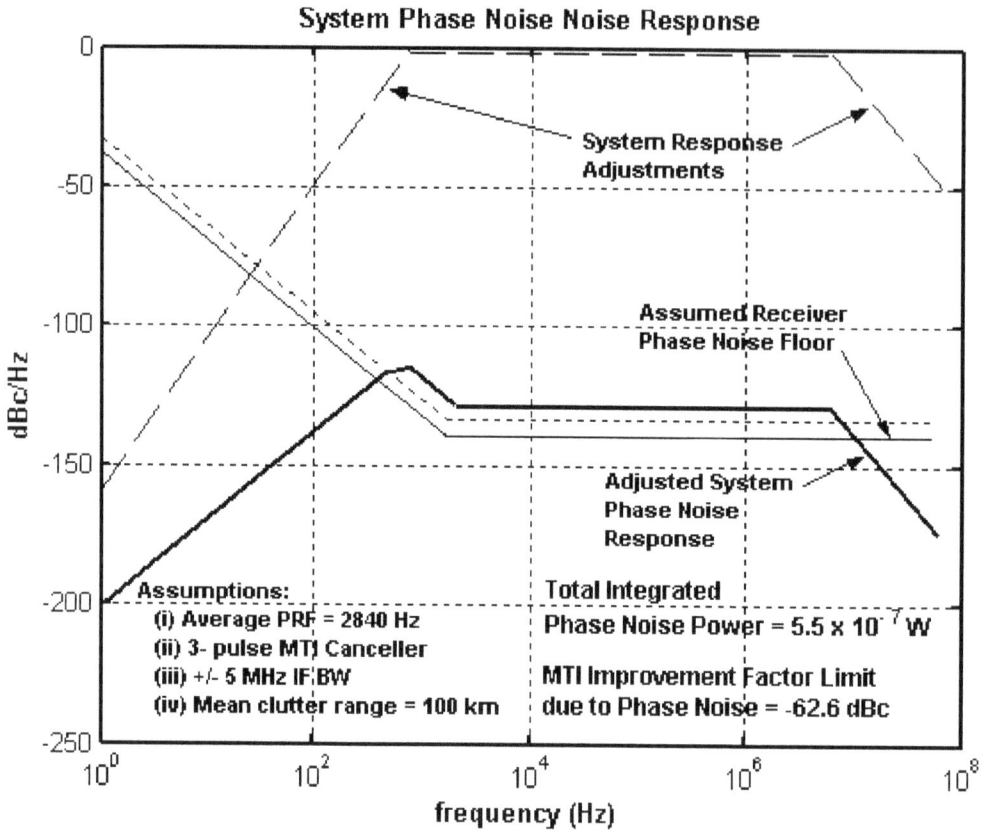

Figure 15.1   System Adjusted Phase Noise Response

modulation waveforms, the antenna mainlobe linearly scans with increasing or decreasing frequency (i.e., depending on whether an "up-chirp" or "down-chirp" waveform is employed by the radar) across the target. Hence, a "scalloping" effect incurs an average loss in target return amplitude.

The problem is to evaluate how small a subarray needs to be to scan a 2,000 MHz waveform out to 60 degrees for a unity figure of merit (*FoM*). The relationship between the figure of merit, subarray width, maximum electronic scan angle, and pulse length is given by:

$$FoM = \frac{(subarray\ width)\sin\left(\theta_{max\ scan}\right)B}{c},$$

(15.11)

where B and c are the waveform bandwidth and the speed of light, respectively. Note that when the *FoM* is approximately unity, about 1 dB of dispersion loss occurs at maximum electronic scan. Solving equation (15.17) for the subarray width yields:

$$subarray\ width = \frac{FoM\ c}{B\ \sin\left(\theta_{max\ scan}\right)}.$$
(15.12)

For the above values, the required subarray width is 0.173 meter.

Therefore if a phased-array uses an antenna aperture of 10 meters^2, then to operate with a 2 GHz LFM waveform at a maximum electronic scan of 60 degrees and to incur only about 1 dB of dispersion loss, then the number of square subarrays required would be approximately:

$$N_{subarrays} = \frac{10}{(0.173)^2} = 334.$$

For less than 1 dB dispersion loss, a larger number of subarrays would be required.

15.3 REFERENCES

[1] J. V. Candy, *Signal Processing—The Modern Approach*, McGraw-Hill,
[2] S. Haykin & A. Steinhardt, *Adaptive Radar Detection and Estimation*, Wiley, 1992
[3] S. Haykin, *Adaptive Radar Signal Processing*, Wiley-Interscience, 2006
[4] S. Kay, *Modern Spectral Estimation: Theory and Application*, Prentice-Hall, 1999
[5] D. Manolakis, *Statistical and Adaptive Signal Processing*, Artech House, 2005
[6] S. L. Marple, *Digital Spectral Analysis with Applications*, Prentice-Hall, 1987
[7] R. A. Monzingo & T. M. Miller, *Introduction to Adaptive Arrays*, SciTech, 2003
[8] R. Nitzberg, *Radar Signal Processing and Adaptive Systems*, 2nd Edition, Artech House, 1999
[9] A. Oppenheim & R. Shafer, *Digital Signal Processing*, Prentice-Hall, 1975
[10] A. Papoulis, *Probability, Random Variables, and Stochastic Processes*, McGraw-Hill, 1965
[11] A. Papoulis, *Signal Analysis*, McGraw-Hill, 1977
[12] H. Van Trees, *Detection, Estimation and Modulation Theory, Part 1*, Wiley-Interscience, 2001
[13] Y. Bar-Shalom, *Multitarget-Multisensor Tracking: Principles and Techniques*, YBS, 1995

[14] Y. Bar-Shalom, *Multitarget/Multisensor Tracking: Applications and Advances,* Artech House, 2000

[15] S. Blackman & R. Popoli, *Design and Analysis of Modern Tracking Systems,* Artech House, 1999

[16] R. Duda, et al., *Pattern Classification,* 2nd Edition, Wiley-Interscience, 2000

[17] K. Fukunaga, *Introduction to Statistical Pattern Recognition,* 2nd Edition, Academic Press, 1990

[18] S. Theodoridis & K. Koutroumbas, *Pattern Recognition,* 2nd Edition, Academic Press, 2003

[19] M. Skolnik, *Radar Handbook,* 2nd Edition, McGraw-Hill, 1990

Index

About the Author

Tom Jeffrey is a Senior Engineering Fellow at the Raytheon Integrated Defense Systems (IDS) business unit. He has over thirty years of broad experience in radar systems engineering covering all phases of design and development, including the development of initial concepts and overall system requirements, system architectures, hardware and software subsystem requirements, detailed algorithms, and system integration and test. He has led systems engineering teams, as well as developed systems engineering training and serves as a mentor to new systems engineers.

Most recently at Raytheon he has served as the system architect for the Cobra Judy Replacement (CJR) radar program and as a consultant on various missile defense radars for both terminal and forward-based ballistic missile defense systems. He served as the Technical Director of the Navy High Power Discrimination (HPD) radar, a shipboard X-band tactical ballistic missile defense radar. For the ground based radar (GBR) family of X-band radars, Tom was responsible for allocating system requirements to hardware and software subsystems, and led a team of systems engineers to develop detailed software requirements for both the tactical and strategic versions of the new ballistic missile defense radars, which evolved into today's THAAD tactical and SBX strategic missile defense radars. He also performed the conceptual and detailed development of air breathing target classification, discrimination, and identification algorithms for the Patriot radar.

Tom has taught a number of radar-related courses at Raytheon on basic and advanced radar, adaptive processing, and architecting methods, and has served as an instructor the Systems Engineering Technical Development Program (SEtdp) for IDS. He is a Senior Member of the IEEE and a member of INCOSE. Tom has authored more than a dozen papers on various radar-related topics. He received his BSEE from the University of Connecticut and the MSEE from Syracuse University. Tom's hobbies include playing and collecting guitars, writing songs, hiking, bicycling, and running. He, his wife, Marie, and Golden Retriever, Tilly, live in Sudbury, Massachusetts.

Radar Range Equations

λ

Compressed Pulse τ_c

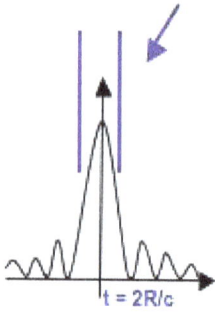

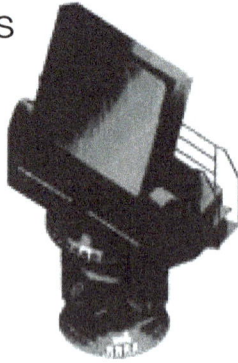

Uncompressed Pulse τ

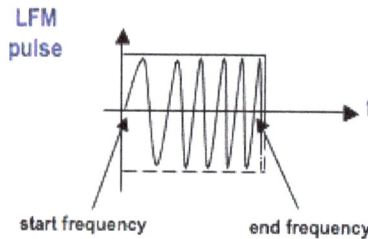

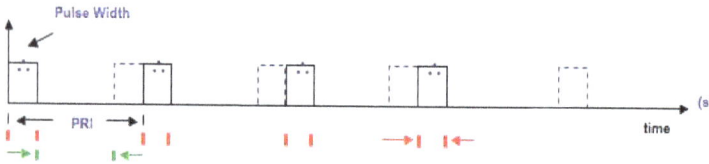

LFM pulse

start frequency   end frequency

t = 2R/c

Pulse Width

PRI

time (s)

Volume Search (Scanning)	$SNR = \dfrac{\sigma \, T_{scan}}{(4\pi) k \, T_S \, R^4 \, \psi_S \, L} \, P_{AVE} \, A_r$
Horizon Search Fence	$SNR = \dfrac{\sigma}{(2\sqrt{\pi}) k \, T_S \, R^3 \, \psi_S \, N \, v_T \, L} \, \dfrac{P_{AVE} \, A_r}{\sqrt{G_r}}$
Track Sensitivity (Staring)	$SNR = \dfrac{\sigma}{(4\pi R^2)^2 \, k \, T_S \, B \, L} \, P_t \, G_t \, A_r$
Track Accuracy	$\sigma_\theta^2 = \dfrac{(4\pi)^3}{2 k_m \, T_t} \left[\dfrac{P_t \, A_r \, G_t \, G_r \, \sigma}{k \, T_S \, R^4 \, L} \right]^{-1}$
Surface Clutter	$SCR = \left(\dfrac{A_r}{A_r'}\right)^2 \left(\dfrac{R_C^3}{R^4}\right) \dfrac{\sigma}{\sigma^\circ \left(\dfrac{c\tau}{2}\right) \tan\phi \, \theta_{AZ}}$
Volume Clutter	$SCR = \left(\dfrac{A_r}{A_r'}\right)^2 \left(\dfrac{R_C^2}{R^4}\right) \dfrac{\sigma}{\sigma^\circ \left(\dfrac{c\tau}{2}\right) \theta_{AZ} \, \theta_{EL}}$

A	Antenna Aperture
AESA	Active Electronically Scanned Array
AESLA	Active Electronically Scanned Lens Array
B	Bandwidth
CDI	Classification, Descrimination & Identification
CPI INTERVAL	Coherent Processing
CW	Continuous Wave
δ	Resolution
FM	Frequency Modulation
G	Gain
k	Boltzmann's Constant
k_m	Monopulse Slope
L	Loss
LFM	Linear FM
MHT	Multiple Hypothesis Tracker
MMW	Millimeter Wave
P	Transmitter Power
P_D	Probability of Detection
P_{FA}	Probability of False Alarm
φ	Depression Angle
PRF	Pulse Repetition Frequency
PRI	Pulse Repetition Interval
R	Slant Range
SAR	Synthetic Aperture Radar
SCR	Signal-to-Clutter Ratio
SCV	Sub-Clutter Visibility
σ	Radar Cross Section
σ_θ	Angle Accuracy
σ°	Clutter Coefficient
SNR	Signal-to-Noise Ratio
STAP	Space-Time Adaptive Processing
θ_3	Antenna Beamwidth
τ	Pulse Length
T_A	SAR Array Time
Tscan	Search Time Frame

Basic Radar Concepts

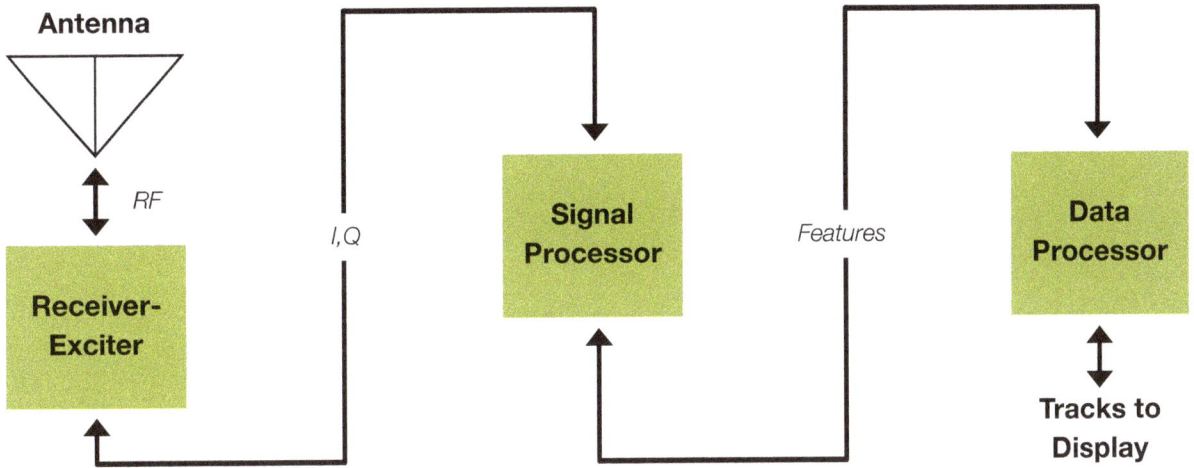

Antenna

```
Receiver-      I,Q      Signal          Features      Data
Exciter    --------->   Processor   ------------->    Processor
   ^ RF                                                  ^
                                                   Tracks to
                                                    Display
```

Radar Range	$R_{min} = \dfrac{c\,\tau}{2} \qquad R_{max} = c\left(\dfrac{T-\tau}{2}\right)$
Decibels	$dB_X = 10\,\log_{10}\left(\dfrac{X_2}{X_1}\right)$
Resolution	$\delta_R = \dfrac{c}{2B} \quad \delta_f = \dfrac{1}{CPI} \quad T_A = \dfrac{\lambda R}{2W_A v \cos\theta_{SQ}}$ $\qquad \delta_\theta = \theta_3 \qquad$ Resolution Cell $\qquad \dfrac{c}{2B}$
Antenna	$\theta_3 = \dfrac{0.9\,\lambda}{L} \qquad \theta_{nulls} = \dfrac{\pm n\lambda}{2L} \qquad \sigma_\theta = \dfrac{\theta_3}{k_m\sqrt{2\,SNR}}$
Frequency	$f = \dfrac{c}{\lambda} \quad B = \dfrac{1}{\tau} \quad PRF = \dfrac{1}{PRI} \quad f_D = -\dfrac{2R_{dot}}{\lambda}$ *SAR Azimuth:* $\quad W_A = \dfrac{\lambda R}{2T_A v \cos\theta_{SQ}}$
Constants	$c = 3 \times 10^8\ m/s \qquad k = 1.38 \times 10^{-23}\ W/Hz-K^\circ$ *SAR:* $\quad R = \left[\dfrac{P_{AVG} A_e^2 W_R \sigma_0}{8\pi\,\lambda\,kT_S v L}\right]^{1/3}$
Detection Swerling I	$P_D = (P_{FA})^{\frac{1}{1+SNR}} \qquad SNR = \dfrac{\ln P_{FA}}{\ln P_D} - 1 \qquad P_{Dcum} = 1 - (1 - P_{D1})^N$
Low PRF	500 Hz to 5 kHz · Range Unambiquous
Medium PRF	10 kHz to 20 kHz · Range & Doppler Ambiguous
High PRF	100 kHz to 200 kHz · Doppler Unambiguous

www.ingramcontent.com/pod-product-compliance
Lightning Source LLC
Chambersburg PA
CBHW061930190326
41458CB00009B/2703